Gender Inequalities

Gender Inequalities
GIS Approaches to Gender Analysis

Edited by
Esra Ozdenerol

CRC Press
Taylor & Francis Group
Boca Raton London New York

CRC Press is an imprint of the
Taylor & Francis Group, an **informa** business

Book cover design by Joanna Göbel

First edition published 2021
by CRC Press
6000 Broken Sound Parkway NW, Suite 300, Boca Raton, FL 33487-2742

and by CRC Press
2 Park Square, Milton Park, Abingdon, Oxon, OX14 4RN

ISBN 13: 978-0-367-18473-5 (hbk)
ISBN 13: 978-0-367-69664-1 (pbk)
ISBN 13: 978-0-429-19658-4 (ebk)

Typeset in Palatino
by Deanta Global Publishing Services, Chennai, India

To my dad for my mother

This book is dedicated to my dad for my mother, and all the

women in the world who deserve a loving partner …

Contents

Foreword

When I began writing this foreword from my hometown of Seattle, not only were we in the midst of a global pandemic, but we were also experiencing one of the worst air quality in the world due to smoke from a convergence of wildfires raging to the south and the east. In a desperate search for information about the safety of returning to outdoor activities, I had become fixated on a mapping service that reports air quality in real time. My obsessive checking showed persistent red dots of "hazardous" air quality glowing from Oakland to Vancouver, B.C. When the much-anticipated marine rains moved in from offshore, I watched in fascination as one by one the cities and towns along the U.S. West Coast shifted from red to orange to yellow and, finally, to green, indicating that it was safe to breathe again.

Being able to track the air quality up and down the West Coast not only gave me information about my own situation; this birds' eye view of the effects of seasonal wildfires was also a stark revelation of the realities of global climate change and put me in solidarity with millions of others experiencing its effects. *Gender Inequalities: GIS Approaches to Gender Analysis* introduces readers to the potential of geographic information systems (GIS) in mapping and measuring gender dynamics across the globe with particular attention to distributions of inequalities. Much like tracking regional air quality, a GIS approach to gender disparities gives us a birds' eye view of patterns regionally and globally. Grounded in the science of geography, GIS technology makes big data accessible and provides place-based analysis: we tend to comprehend pictures and maps better than charts and graphs. GIS allows us to visualize, track, and make sense of everyday patterns and phenomena, including air quality, coronavirus outbreaks, and more mundane features such as the location of coffee shops in your neighborhood. For instance, natural scientists use GIS technologies to track species migrations and habitats, which are closely linked to the overall health of the planet and, by implication, our own lives. In the social sciences, GIS provides visual maps of social phenomena such as crime distributions, poverty density, employment patterns, and food and health care access.

As a research tool, GIS provides a "big picture" and enables social scientists to contextualize their experiences by grounding them in space and place. This visual mapping brings experiences and patterns into focus in ways that make them more relatable. For instance, in a 2008 paper published in *Social and Cultural Geography*, feminist geographer Mei-Po Kwan demonstrated the power of GIS by translating oral histories (a typical feminist research approach) into visual narratives. She mapped the daily experiences of Muslim women in New York City immediately following the September 11 (2001) attacks on the World Trade Center, illuminating their truncated movements around the city as a result of anti-Muslim hostilities. The resulting "emotional geographies" located the experiences of these women in time and space and significantly broadened our understanding of what it means to be afraid to move about freely.

With *Gender Inequalities,* editor Esra Ozdenerol gives us nine richly crafted case studies that reflect advances in GIS and demonstrate the significant contributions these spatial methodologies offer for feminist social science. The scholars featured here are using innovative GIS approaches to provide place-based analyses that zoom in on local patterns of gender difference and zoom out to reveal persistent large-scale inequalities. As a sociologist, I first came to appreciate the significance of geographical place-based analyses when I had the opportunity to edit a two-volume *Encyclopedia of Gender and Society* (Sage, 2008). Published more than a decade ago, the 500+ entries on various aspects of gendered experiences across the globe suggested a need to be able to view these dynamics as part of larger systems of inequalities. Public health scholars, development researchers, and environmental justice scholars have been among the first to appreciate the application of the theories and methods of geography. Geography taught us the importance of space and place (it seems odd to recall a time when we weren't so fully aware of the need for grounded analyses). Feminist geography has provided conceptual and methodological tools to map both local dynamics of inequality and larger persistent patterns. *Gender Inequalities* joins the ranks of this hugely important scholarship and moves it forward by demonstrating the potential of GIS technologies for spatial analysis. The cases represented here include analyses of employment sectors, migrant laborers exposure to pollution, and distributions of opioid use and deaths, to name just a few. In each case, the use of GIS analysis reveals patterns that are non-intuitive and may surprise the reader. There is a lot to learn here about a wide range of topics.

Gender Inequalities also represents a group of scholars who are recontextualizing earlier critiques of GIS. In this regard, the book is notable for its contributions not only to a geography of gender inequalities but for its methodological implications. When GIS technologies first began to appear in the social sciences, feminist scholars were quick to pounce on supposed epistemological differences. Citing GIS as grounded in "masculinist" and "positivist" frameworks, feminist critiques in the 1990s raised concerns about claims to objectivity and neutrality. They raised objections against positivist, big data research as aiming to generalize and to seek universal principles. Big data, so it was argued, eclipses the subject and rejects the reflexivity that is central to feminist logics. Further, big data was conflated with notions of "Big Brother" and came to be seen as a powerful surveillance tool used by the state to oversee and control the lives of individuals. To the extent that geospatial technologies were and still are being used by the military for surveillance and by commercial enterprises to mine data on individual consumer habits, these fears are grounded. But early in the twenty-first century, a few feminist scholars began to question the inherent positivism of GIS and wondered, collectively, if it might be developed as a tool for critical social research and progressive change.

In 2002, feminist geographer Nadine Schuurman noted that GIS technology "is only as masculinist as we allow it to be." She challenged feminist scholars to consider ways in which the technologies could be informed by and compatible with feminist perspectives. Scholars such as Sarah McLafferty and Mei-Po Kwan took up the challenge, claiming that GIS methods have the potential to convey fine spatial scale and complex levels of detail that more fully represent the socio-spatial contexts of women's lives. Taking a cue from Donna Haraway's philosophy on feminist subversions of the master–subject dialectic, these scholars posited that GIS didn't have to be a "god's-eye" view but rather could be used for visual practices based in feminist reflexivity. Mei-Po Kwan's mapping of the daily movements of Muslim women post 9/11 is an example of GIS informed analysis centered in a view from the body.

In seeking greater compatibility between feminist research and GIS technologies, contemporary feminist geographers are also looking for ways to integrate qualitative data in GIS projects. To this end, an emerging qualitative, critical GIS is reclaiming GIS from "technological determinism" through the development of localized data bases and dedicated algorithms (rather than pre-made general data sets). *Gender Inequalities* offers exciting examples of these feminist-informed, contemporary applications of GIS spatial mapping. The entire project inspires me.

As a long-time feminist sociologist teaching research methods, I frequently find myself telling students that it's all in how we frame the questions. Grounded, contextualized maps provide not only detail but a visual representation of patterns and connections that enables us to ask grounded questions. Understanding dynamics and tracing patterns is key to doing progressive social research aimed at making a difference. The cases represented in *Gender Inequalities* invite new considerations about the relationship between theory and empirical research. I'm excited to see feminist scholars not only joining these conversations but, as evidenced by this collection of case studies, mapping the horizons and charting (literally) the way forward.

Jodi O'Brien
Seattle, 2020

Preface

My interest in writing this book stems from the quest to investigate the magnitude and spatial distribution of gender inequality and how it is manifested around the world, particularly, when and where it occurs. This spatial and spatial-temporal perspective enhances the role that GIS can serve in both mapping and measuring gender inequalities with consistent methodologies. Adopting GIS and spatial analysis to gender studies and gender inequalities is not new, though there is no recent text addressing this rapidly growing interest. My book fills this gap by exploring new linkages between GIS and gender inequality and research needs.

Gender inequality is entrenched in the cultural, political, and market systems that operate at household, community, and national levels. Overarching global changes in access to markets, climatic conditions, and the availability of natural resources and ecosystem services intensifies disparities in income, assets, and power among genders and justifies the publication of this timely text. To understand these gender dynamics at the macro (e.g. subnational) and micro (e.g., village) levels will require detailed research that GIS and mapping allows. This book is organized to foster this understanding. It involves mapping gender inequalities in their geographic context and draws examples from the United States and the world to identify relevant applications.

Since gender inequalities is a research area that is very wide and with strands into many academic traditions and the international development sphere, this book is aimed at diverse academics and development institutes that have invested heavily in GIS technology and are currently conducting development research with the aid of this powerful tool. It is a unique textbook for geographers, public health practitioners, local development practitioners, sociologists, epidemiologists, criminologists, politicians, economists, environmentalists, GIS scientists, and health and research professionals interested in applying GIS and spatial analysis to the study of gender inequalities.

The main questions I seek to answer for the reader are: How do I analyze gender inequalities using GIS? Which spatial approaches and indicators do I use to map and measure gender inequalities? How do gender inequalities vary over time and space?

This book consists of ten chapters of selected papers presenting in-depth case studies that adopt GIS and spatial analysis for visualizing and mapping gender inequalities of various regions and countries in the world. These global case studies provided in each chapter explore the world of gender inequalities and get directly involved with some of the GIS and mapping applications.

Chapter 1 provides an overview of the current role of GIS in the context of gender inequalities. The purpose of this chapter is twofold: first, the chapter is intended to present how GIS can be integrated in gender inequalities. It argues that GIS scientists can contribute to research on gender inequalities by teasing out the connections between place/space and gender inequalities and by utilizing GIS tools and spatial methodologies to explore these associations. Inventive new tools for analysis and big data and open data platforms provide a strong foundation for innovative place-based gender inequalities research. In summary, GIS and spatial methodologies are greatly advancing how place characteristics are measured and how their associations with gender inequalities are assessed. Those of us concerned with gender and development issues have much to gain from the use of GIS, particularly if we are aware of its ability to empower women and the poor, and different gender experiences such as LGBT communities.

Second, the chapter is an introduction to the global problem in gender inequality. It covers a detailed description of gender inequality from a global perspective, how it still exists globally despite substantial national and international measures that have been taken toward gender equality. What are the evident progresses, and the alarming issues regarding gender inequalities that still prevail today? What makes gender inequality a global priority as a fundamental step in both human development and economic progress? This part of the chapter illustrates global and country-level maps of measures of gender inequalities, such as access to basic education, health and life expectancy, equality of economic opportunity, and political empowerment.

Chapter 2 investigates the role that GIS and geography can serve in both understanding and eliminating gender-based violence, especially, how research on domestic violence (DV; i.e., violence that occurs between romantic and marital partners) has recently begun to incorporate geographic information systems (GIS) technology in order to better understand the nature of this form of violence, and how crime mapping can be adapted for the criminal justice response to DV, including how it can: (1) improve law enforcement response; (2) increase victim reporting; and (3) increase victim access to services.

Chapter 3 discusses applying GIS and spatial analysis to the prevalence and incidence mapping of intimate partner violence (IPV) and geospatial factors that influence help-seeking and resource availability. First, it introduces a literature that has identified numerous geographic correlates of IPV, including concentrated disadvantage and perceived disorder, and buffering effects of social cohesion and support. Limitations of the literature include the use of potentially biased self-report measures and arbitrary approaches to quantifying environments (e.g., relying on census tracts). It concludes that GIS has proven to be a useful tool for studying IPV that may alleviate concerns regarding subjective measurement and artificial boundaries.

Chapter 4 discusses the spatial disparity of gender-representation across industry types in the United States. This analysis finds that women are far less represented in almost all types of industries, even though significant gains have occurred in many traditionally male-dominated industries, such as primary and secondary sectors, as well as in professional/scientific and other service industries. The GIS maps illustrate wide gender gaps in industry-based location quotients across U.S. counties. This analysis provides insights into a critical need for place-based policies that can create opportunities for including women in typical male-dominated industries.

Chapter 5 explores the social and environmental injustice experienced by female migrant workers at Guiyu town, Guangdong province, China, in the context of both environmental pollution and governance. Guiyu town has been highly polluted by the electronic and electric waste recycling industry and is now experiencing rigorous environmental governance. Using a qualitative GIS approach that can visualize the ethnographic and other qualitative data with GIS applications, it reveals the various ways in which power relations based on class, migrant status, gender, and age intersect to produce social and environmental injustice. It shows that female migrant workers are particularly vulnerable to the effects of local environmental pollution and degradation in both working and living spaces. This chapter also indicates that GIS and story maps can be used to bridge the gap between qualitative and quantitative analyses that often results in the ignorance of the environmental injustice faced by females. Including spatially referenced pictures (interviews, participant observation, focus groups) archival data, as well as demographic data and other attribute data with "real world" locations in story maps gives researchers and decision-makers a perspective not from space but from the level of the research participant. This helps the user realize that the phenomenon under study occurs at some real, specific, and complex location.

Chapter 6 presents a social vulnerability index to identify spatial patterns of social vulnerability and gender inequalities among Mexican households. Using geo-spatial analysis, households were classified into eight different types based on gender of the head of the household and the presence or absence of children. Results showed that female-headed households with children are the most vulnerable group. The spatial clusters of high social vulnerability and gender inequalities are mainly distributed in the southern region of Mexico, in the states of Oaxaca, Chiapas, and Guerrero. These entities are well known for having the highest level of poverty and unregistered employment in the country, together with an extensive presence of indigenous population.

Chapter 7 presents the United States' opioid crisis over the past two decades. The nation is currently experiencing the third wave of the crisis, which has seen a rise in mortality from synthetic opioids. These mortalities are analyzed in relation to the decedents' gender, race, age, and urbanicity, as well as the deaths' spatial and historical characteristics. The findings challenge the notion that the crisis mostly affected male, white, middle-aged, middle-class, rural, and suburban users and show the importance of considering gender, race, and space in developing medical treatments, health interventions, and public policies in response to the drug crisis.

Chapter 8 discusses the commitment to "leave no one behind" as the heart of the 2030 Agenda for Sustainable Development, adopted by all United Nations member states in 2015. The first step in operationalizing this principle is identifying who are the most marginalized and how they fare on key markers of well-being, particularly in comparison to other groups in society.

In this chapter, 2017 GIS data from the Pakistan Demographic and Health Survey is used to identify inequalities among women and girls. Mapping multiple deprivations that are spatial in nature reveals that being in a remote rural area or urban slum in Pakistan is associated with poverty, unavailability of improved water and/or sanitation facilities, and a lack of access to health facilities. The multidimensional maps derived from this analysis put into sharp focus the tendency of deprivations to cluster together and bolster the call for an interdependent approach to the 2030 Agenda and its 17 development goals.

Chapter 9 discusses the long-standing challenges in establishing gender parity in the transportation workforce in the United States. Mapping gender inequality in this regard is further complicated as transformational technologies and socio-economic trends redefine the transportation sector into the more representative new mobility workforce, which includes all of the disciplines required to design, develop, operate, and maintain the mobility systems that transport people and goods. Using GIS tools, a layered system model is presented to document and visualize new mobility workforce data that include oversight, transactional, and operational levels, while also encompassing governmental, industry, and educational sectors of the economy. This approach provides visually accessible data-driven frameworks for developing benchmarks that comprehensively define the new mobility workforce and create a platform for policymakers to better visualize workforce gaps and target investments to address them. The chapter also assesses data needs and provides recommendations for developing longitudinal studies to track progress.

Chapter 10 presents a study that utilizes geospatial statistical tools and state-level admission data to examine gender inequalities in higher-education enrollment in Nigeria and to investigate the key factors on enrollment. For gender inequality analysis, percentage female enrollments (PFE) are computed. Spatial clusters in PFE are assessed using global Moran's *I* while determinants of PFE are assessed with geographical weighted regression (GWR). Results indicate a general decrease in PFE from states in the south to states in the north throughout the study years: 2005–2015. The finding from GWR suggests that the persistent regional variations in female enrollment over time can be explained by the size and diversity of applicant pool. The study underpins the need for affirmative programmatic efforts to promote female enrollment in higher education in Nigeria.

The global case studies provided in each chapter explore the world of gender inequalities and get directly involved with some of the GIS and mapping applications. A notable feature of "spatial gender inequalities" is its attempt to fill the gap by exploring new linkages between GIS and gender inequality and research needs. That is why I wrote this book: to share the stories of people, communities, and countries that have given focus and shown urgency toward eliminating gender inequalities using GIS tools. Getting around barriers of poverty, distance, ignorance, doubt, stigma, religious, and gender bias only happens by understanding the meaning and beliefs behind a community's and/or country's practices in the context of their values and concerns.

As the seventh Secretary-General of the United Nations, Koffi Annan, states, "Gender equality is more than a goal. It is a precondition for meeting the challenge of reducing poverty, promoting sustainable development and building governance." I hope that I have conveyed the amazing breadth of GIS use in studying gender inequalities with this book. Our eagerness to learn and adopt these technologies and our optimism for the world and our work together are among the greatest sustaining forces toward eliminating gender inequalities.

Esra Ozdenerol
Memphis, TN

Acknowledgments

First and foremost, I would like to thank the contributors of this book from all over the world for the provision of their research, stories, and maps. It is a privilege to share their work with the world. They weave together vulnerable, brave storytelling and compelling data from their countries. They provide an unforgettable narrative backed by startling GIS data as they present the issues that most need our attention – from intimate partner violence to lack of access to education to gender inequity in the workplace. They also demonstrate that gender equality without connection misses the whole point. People can be equal but still be isolated. I cannot win if you lose. If either of us suffers, we suffer together. GIS technology blurs the borders between human beings and empowers them on the road to gender equality. Gender equality can empower women, and empowered women will change the world.

I owe a great deal to the many people who have provided inspiration, help, and support for what would eventually become this book: Irma Britton, Senior Editor, and Rebecca Pringle, Project Coordinator of CRC Press, for their patience and guidance throughout this project. I thank Joanna Gobel for coming up with a fantastic cover design. I owe my perpetual gratitude to Dr. Jodi O'Brien, Professor of Sociology at the Seattle University, who wrote the foreword to this book. I thank my students: Dr. Ryan Hanson for coauthoring a chapter with me and Jacob Seboly and Faria Naz Urmy who contributed to the development of the maps, editing, and formatting. I offer very special thanks to our department administrative assistant, Jessica Albernathy, for her contagious smile.

There are a number of women and men I've been lucky enough to meet in my life who have taught me truths that will last a lifetime – my teachers at Ogretmen Kubilay Elementary School in Yenimahalle and Ankara Ataturk Anatolia High School in Ankara, Turkey, my professors at Ankara University, Middle East Technical University, Louisiana State University; and my mentors and role models in creating change in the world.

I would also like to thank all my friends who have lent their support and listened to my complaints for the past years. Special thanks go to Phuong Nguyen, Lyn Wright, Gregory Taff, Ashley Fowler, Johanna Leibe Gamard, Bunny Williams, Nazan Gursakal, Asli Yalcin, and Funda Cam. I would like to thank Maxine Smith STEAM Academy and University Middle School teachers for their commitment and dedication to my sons' education which made my life easier but motivated me to be a better parent and an educator.

My dad, Erdogan Ozdenerol, who has served as the Court of Appeal Judge of Turkey, has been so supportive of me and my career in every imaginable way, believing so strongly in me and in women and offering constant encouragement to me as I made the decision to become a public and academic advocate. He is the mentor of a lifetime, and I can never thank him enough.

I want to thank my mother, Umran Ozdenerol, who is the greatest Math teacher in the world, who inspired the love and values that started me in this work in the first place. I would like to thank my parents, who gave me a childhood steeped in the deep values of love; my sister, Emel Ozbay, who shares love and laughter; her husband, Alp Ozbay, and my beautiful and intelligent nieces, Irem and Zeynep Ozbay; my children, Derin and Deniz, who constantly inspire me to grow.

I owe a debt of gratitude to many people I met around the world, especially the women (and men) who welcomed me into their homes and communities during my travels, told me their dreams, and taught me about their lives. Thank you from the bottom of my heart.

I wrote this book on gender inequalities to contribute to the #MeToo movement in a meaningful way. This book is dedicated to the victims of intimate partner violence in the world, women who faced constant violence from their partners who were poor and marginalized. The #MeToo movement went global. It has awakened us to the ongoing savagery inflicted on our daughters, sisters, and mothers. Women and girls globally continue to be discriminated against, manipulated, and oppressed. Even in Western nations, women are now speaking out about how they were discriminated and sexually exploited. Even in Western nations, relatively few women serve in legislative assemblies, and seldom are they ever represented in supreme judicial bodies. As our world is going through a precarious time when human rights, freedom, and peace are under assault, resulting in human rights abuses, arms proliferation, proxy wars, and waves of refugees, we should all stand up for women's protection and their rights.

I believe in people all along the gender spectrum who have every right to wear their wounds openly and are working to end gender inequalities and give voice to their experiences on the road to equality. The voices shared are voices that matter and demand to be heard.

Finally, I want to thank all the enthusiastic supporters of GIS. After teaching GIS for over 20 years, I know that you – like my own students – will enjoy this subject and its substantial potential to end gender inequality and put gender on the map. Go to GI-Yes!

Esra Ozdenerol
Memphis, TN

Editor Biography

Dr. Esra Ozdenerol (Ph.D., 2000, Louisiana State University) is Dunavant Professor of Geography at the Department of Earth Sciences and the Director of Graduate Certificate in Geographic Information Systems (GIS) at the University of Memphis. She also serves as Adjunct Professor of Preventive Medicine and Health outcomes Policy at the University of Tennessee Health Science Center and Adjunct Professor of Biology at Arkansas State University.

Dr. Ozdenerol has been with the University of Memphis since 2003, where she teaches GIS and its applications to health disparities. Her research interests entail the use of geospatial technologies in a diverse range of environmental health issues, gender and health inequalities: Covid-19 pandemic mapping, birth health outcomes, opium epidemic, infectious disease epidemiology, children's lead poisoning, etc. She served as Associate Director of Benjamin L. Hooks Institute for social change between 2010 and 2013. In 2010, she served as President of the Memphis Area Geographic Information Council, a non-profit organization of GIS professionals. She was invited by the National Academies of Science on the Contribution of Remote Sensing for Decisions about Human Welfare and was also recognized as a "Health Research Fellow" by the University of Memphis. She launched the first interactive mapping website featuring the pivotal events of the civil rights movement. Her textbook *Spatial Health Inequalities: Adapting GIS Tools and Data Analysis* has been recognized by leading institutions such as AAG and adopted at various geography and public health departments. She has lectured nationally and internationally on GIS and Remote Sensing and issues of health disparities. With her key note speech, she opened the Refugee Week at Refugee Therapy Center in London in 2013. She has developed a GIS workshop series for professionals and campus community that addresses and supports different aspects of the data life cycle and data used with different GIS software and tools. Dr. Ozdenerol has visited 50 countries, lived in three continents, and brings her travels into her World Geography and GIS courses. She has published bilingual children's books for teaching languages and changing attitudes about other cultures and geographies. Dr. Ozdenerol is the lead singer of Memphis City Sound Chorus, an award-winning women's chorus of Sweet Adeline's International. She has an unstoppable passion for singing harmony as well as teaching geography.

List of Contributors

Roberto Ariel Abeldaño Zuñiga
PostGraduate Department
University of Sierra Sur
Oaxaca, Mexico

Olayinka A. Ajala
Department of Geography
Obafemi Awolowo University Ile-Ife
Ile-Ife, Nigeria

Adesina A. Akinjokun
Department of Architecture
Obafemi Awolowo University Ile-Ife
Ile-Ife, Nigeria

Akanni I. Akinyemi
Department of Demography and
   Social Statistics
Obafemi Awolowo University Ile-Ife
Ile-Ife, Nigeria

Ginette Azcona
Research and Data Section
UN Women
New York, NY

David O. Baloye
Department of Geography
Obafemi Awolowo University Ile-Ife
Ile-Ife, Nigeria

J. Gayle Beck
Department of Psychology
University of Memphis
Memphis, TN

Antra Bhatt
Research and Data Section
UN Women
New York, NY

Javiera Fanta Garrido
Instituto de Investigaciones Gino
   Germani
University of Buenos Aires
National Council of Scientific and
   Technical Research
Buenos Aires, Argentina

Ryan Baxter Hanson
City of Memphis
Division of Housing and
   Community Development
Memphis, TN

Amaia Iratzoqui
Department of Criminology and
   Criminal Justice
University of Memphis
Memphis, TN

Stephanie Ivey
Herff College of Engineering
University of Memphis
Memphis, TN

Angela D. Madden
Public Safety Institute
University of Memphis
Memphis, TN

James C. McCutcheon
Department of Criminology and
   Criminal Justice
University of Memphis
Memphis, TN

Benjamin Olson
Center for International Trade and
   Transportation
California State University Long Beach
Long Beach, CA

Moses O. Olawole
Department of Geography
Obafemi Awolowo University Ile-Ife
Ile-Ife, Nigeria

Esra Ozdenerol
Department of Earth Sciences
University of Memphis
Memphis, TN

Esra Ozdenerol
Spatial Analysis and Geographic
Education Laboratory
Department of Earth Sciences
University of Memphis
Memphis, TN

Alison M. Pickover
Maven Clinic
New York, NY

Tyler Reeb
Center for International Trade and
Transportation
California State University Long
Beach
Long Beach, CA

Madhuri Sharma
Department of Geography
University of Tennessee
Knoxville, TN

Ye Zhang
Faculty of Education and Arts
Australian Catholic University
Sydney, Australia

1

The Global Problem in Gender Inequality: Putting Gender on the Map with GIS

Esra Ozdenerol

CONTENTS

This chapter introduces gender in Geographic Information Systems (GIS) and explains the substantial potential that GIS holds for supporting efforts to end gender inequality and put gender on the map. Though the role of GIS in gender study is new, GIS can serve to illustrate distribution and access to resources, which can contribute toward the solution of socio-economic issues and the reduction of disparities between men and women while also advancing LGBTQ rights. GIS has characteristics inherent in its structure such as data collection and representation that make it possible to produce gendered knowledge. By definition, GIS is a framework for gathering, managing, and analyzing data (Cromley 2003). Rooted in the science of geography, GIS integrates many types of data. It analyzes spatial locations and organizes layers of information into visualizations using maps and scenes. With this unique capability, GIS can reveal deeper insights into gender data, such as patterns, relationships, and situations. This process starts with the research questions we ask, moves into the data we collect, and is finalized in the graphic representations we produce. While there is currently a lack of gender data readily available for spatial analysis, awareness of the issue and demand for data will bring necessary information forward with the goal of putting gender on the map. GIS scientists contribute to gender inequality research by teasing out the connections between place and inequality and by utilizing GIS tools and spatial methodologies to explore these associations. Inventive new tools for analysis, big data, and open data platforms provide a strong foundation for innovative place-based gender inequalities research. In summary, applications that utilize GIS and spatial methodologies enhance the measurement of place characteristics and the assessment of their associations with gender inequalities.

First and foremost, GIS has the power to aid and influence international development work by combining different datasets into one visual image. Integrating statistical geospatial data with socio-economic information and analysis and mapping indicators of development from district to country level facilitate the assessment of progress and gaps across all Sustainable Development Goals (SDGs) of the 2030 Agenda (UN 2019). A hallmark of the 2030 Agenda is that it applies to all countries, all people, and all segments of society, while promising to prioritize the rights and needs of the most disadvantaged groups, from access to education and health care to clean water and decent work. The spread of information and communications technology and global interconnectedness has great potential to accelerate human progress, to bridge the digital divide, and to develop knowledge societies, as does scientific and technological innovation across areas as diverse as medicine and energy. GIS analysis and mapping of gender inequality indicators reflect the realities of the lives of women and men and describe the roles of women and men in the society, economy, and family. This can be utilized to formulate and monitor policies and plans, monitor changes, and inform the public.

The online GIS platform is used to establish a comprehensive mapping of, and serve as a gateway for, information on existing science, technology, innovation, initiatives, mechanisms, and programs. There is a range of practical online/mobile interactive GIS maps, dashboards, and applications. Some of these dashboards and applications receive data updates in near-real-time. The online GIS platform facilitates access to information, knowledge, and experience, as well as the best practices for science, technology, and innovation facilitation initiatives and policies. The online platform also facilitates the dissemination of relevant open access data generated worldwide. Figure 1.1 is an example of an online GIS dashboard of Covid-19 cases in the context of gender inequalities and women's health care needs in the world. As GIS depicts spatial patterns in gender-related data, issues and trends such as fertility rates, literacy rates, and population density are revealed, and gender inequality index and other development indexes for countries and districts can be mapped. The countries and districts are then ranked in order of most

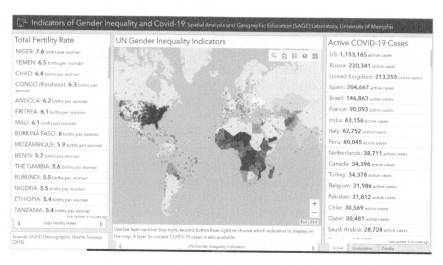

FIGURE 1.1

A screenshot of an online GIS dashboard of Covid-19 and Gender Inequalities (Source: Spatial Analysis and Geographic Education Laboratory, University of Memphis, https://cpgis.maps.a rcgis.com/apps/opsdashboard/index.html#/f4000bb07cfc4ba4989fa72e1d056b6c).

to least developed. With GIS, key issues can be quantified and investigated more effectively. Spatial patterns in socio-economic data at finer scales, such as at the district level, also reveal issues and trends that would otherwise be missed by data aggregation by political or other units.

Since the 1990s, development interventions have been formulated and implemented in an integrated manner using GIS. The later invention of cloud and mobile technology has added district-level digital datasets of development regions and has facilitated wider dissemination of GIS data, technology, and applications. For example, the MENRIS (Mountain Environmental and Natural Resource Information Service) database is developed by a decentralized network of institutions to facilitate environmental and natural resource planning, monitoring, and management among several mountain nations including Nepal, Bhutan, India, Pakistan, China, Bangladesh, and Myanmar. This enlightens the region with new technological innovation in information sciences and enables the collection, storage, and dissemination of key biophysical and socio-economic data at national and district levels. MENRIS case study series are published and distributed to facilitate wider dissemination of GIS technology and applications (Murai and Singh 1998). This project uses a breadth of variables such as literacy rates, population density, agricultural credit, and slope steepness to construct a development index for each district.

An integrated approach to addressing gender inequity and making use of the opportunities for development should have a strong database and an appropriate methodology for analysis. There are many challenges, however, to operationalizing a methodological approach that captures the intersection of different forms of discrimination and inequities. These challenges include data limitations and identifying which forms of discrimination are relevant in each context. Wealth and income-based discrimination are understood to be relevant across countries, but other forms of discrimination are more context-specific per country and/or region in a country. For example, land distribution translates into economic disparity. GIS is used for securing land rights across the world today. In Tanzania, under the Evaluation, Research and Communication (ERC) project, the USAID (United States Agency for International Development) piloted a project to crowd-source land rights information at the village level using mobile technology. The Mobile Application to Secure Tenure (MAST) project supported identified needs of the Government of Tanzania to improve land governance and lower the cost of land certification programs (USAID Land Projects 2016). The mobile platform technology was able to map a community's land of 1,000 plots in four weeks. The immediate result of this was that women went from having no land ownership to boasting of 40% of the land. Another 30% of the land went under joint titling, and the remaining 30% under male ownership. It is a right that people had never had before. Formal land administration systems (LAS) in developing countries have generally not met the need for accessible, cost effective, and appropriately nuanced land registration. As a result, large majorities of rural dwellers (and many urban dwellers) live without formalized rights to land and other valuable resources. This lack of documentation may constrain the ability of individuals and communities to leverage their land-based assets for improved economic outcomes, to limit environmental harms, and to engage in collaborative contracting with prospective investors in land that leads to equitable sharing of benefits. Given the rising concerns related to inappropriate and potentially harmful transfer of land rights from vulnerable populations to domestic and foreign investors, many tenure experts view the need to document existing rights in a participatory and efficient way as a high priority.

Land tenure and land security are a priority in the Sustainable Development Goals. GIS technology is the most cost-effective model that can be utilized to achieve these SDGs considering the magnitude of the problem. More than 90% of the African continent still does not have a formal title or any type of recognized secure tenure to land. GIS will likely play a prime role in finding a solution for this problem in the next 15 years. Figure 1.2 reflects partner countries to a broader initiative led by the Global Donor Working Group on Land to compile information on all donor-funded land and resource governance programs (GDWG 2020). The group, established to coordinate activities among donors and development agencies, strives to improve access to secure land tenure and property rights for people all over the world, with an emphasis on vulnerable groups, including women and indigenous people. GIS provides a hub for information related to land governance, which improves information sharing, coordination and collaboration on land-related programming, and interventions among the member

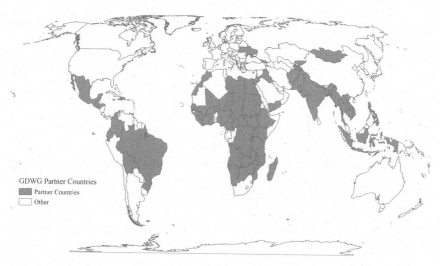

FIGURE 1.2

Partner countries to the Global Land Governance Group Programme (Mapped by Spatial Analysis and Geographic Education Laboratory, University of Memphis, Data source: GLGP–The Global Land Governance Programme Map & Database (2020). Accessed May 5, 2020. https://landgov.donorplatform.org/).

organizations; strengthens partnership among partner countries to improve land governance and transparency; and engages with and supports the private sector to improve land governance by strengthening core business procedures (such as due diligence related to land-based investing). The Global Land Governance Programme Database and Map currently lists information on over 500 programs in more than 100 countries (GLGP 2020).

Reducing the impact of gender inequality and gender norms and expectations on women's and men's health conditions and ensuring healthy lives and promoting well-being for all is one of the SDGs. The global focus has now shifted to the achievement of the SDGs which similarly propose to improve maternal health and reduce mortality to less than 70 per 100,000 live births and neonatal mortality to at least 12 per 1,000 live births by 2030 (UN 2015). Achieving these goals will require national maternal and newborn health (MNH) programs to address underlying localized inequalities (Bhutta and Reddy 2012). GIS applied to maternal and neonatal health data holds substantial potential for supporting efforts to end preventable maternal and newborn deaths. Realizing this potential will require improved access to high-quality MNH data at needed resolutions for decision-makers at multiple levels, increased understanding of and skills in using both the software and the maps for planning and implementing MNH programs, and consistent involvement of the community in the mapping process as well as in the use of high-resolution maps.

Beyond mapping mortality distribution, GIS has largely been used for monitoring and reporting progress of maternal health interventions and quantifying access to facilities and care (Ebener et al. 2015). An example of utilizing GIS for non-mortality mapping includes linking the State of the World's Midwifery findings with GIS to ensure fair distribution of services and priority for the worst off (UNPFA 2014). Researchers and policymakers have been calling for more equitable improvement in maternal and newborn health, specifically addressing hard-to-reach populations at subnational levels. Data visualization using mapping and geospatial analyses plays a significant role in addressing the emerging need for improved spatial investigation at a subnational scale. This correspondence identifies key challenges and recommendations so that GIS may be better applied to maternal health programs in resource-poor settings. The challenges and recommendations are broadly grouped into three categories: ancillary geospatial and MNH data sources, technical and human resources needs, and community participation (Molla et al. 2017). Ancillary data includes population estimates, subnational boundaries, roads and rivers, etc. Health data can be separated into non-routine data, such as those from surveys, and routine data, such as those from health information systems, health facility registries, maternal death surveillance and response systems, and vital registry systems. Whether examining access to health facilities or predicting skilled birth attendance, the local environment plays a critical role in influencing MNH. As such, close attention must be paid to the types of databases used in analysis, as well as geographical division and time of data collection (Molla et al. 2017).

The DIVA-GIS project is a commonly used, consolidated source of country-level and global ancillary data that is freely available (DIVA-GIS 2020). Information on the distribution of populations is also freely available through the WorldPop project (www.worldpop.org), including high-resolution data on human population distributions for countries in Africa, Asia, and Central and South America (Detres et al. 2014). Specifically, in the context of MNH, distributions of live births, pregnancies, and women of childbearing age are available both on the 100-meter level and at the administrative unit 2 level, where applicable, so are important covariates such as poverty, literacy, and urban change. Non-routine survey data, such as those from the Demographic and Health Surveys (DHS), provide a rapid entry into the use of GIS for MNH. DHS data provides users with a readily accessible, freely available source of geo-located household and facility-based surveys which can be used to model an array of MNH outcomes, both within and across countries (DHS 2020). Surveys are often conducted every 5–10 years, not frequently enough for general program monitoring. Additionally, maternal mortality, a key indicator, can usually only be mapped at the national level because the commonly used sisterhood methods do not record the location of deaths for making subnational estimates (Ahmed and Hill 2011). The spatial resolution of surveys has limited their use to the state or province level, although recent guidance on interpolated maps using DHS data may lead to new avenues for detailed spatial investigation (Burgett 2014).

It is known that both the development and the use of high-quality maps should involve community engagement and participation (Detres et al. 2014; Gozdyra et al. 1999). Recent practices of participatory mapping have facilitated monitoring real-time data for mapping, supported interpretation of spatial analysis results, and fostered ownership and decision-making by the communities engaged (Ruktanonchai et al. 2014).

GIS and remote sensing technologies are also used to disrupt the market of labor trafficking through uncovering the location of traffickers and their attempts to transport victims. Satellite imagery is playing a significant role in achieving this. For example, DigitalGlobe, a company that provides high-resolution images of the earth, is able to spot slave ships in the seas (Digital Globe 2020). Using powerful satellites, seas that have long remained lawless can now be policed. DigitalGlobe also investigates brick kilns in India and fisheries on Lake Volta in Ghana, two major industries where child labor exists (Donovan 2017). Geospatial technology can also lead to a way out from trafficking. For instance, a report from the USC Annenberg Center on Communication Leadership and Policy, Technology and Labor Trafficking in a Network Society describes the story of a woman from the Philippines who was stranded in Malaysia and deceived by traffickers (Latonero et al. 2015). She was thrown in prison and interrogated, but the Philippine government was able to intervene and help her because she had hidden a phone in her jail cell. Based on this Philippines case study and analysis of supply chains, online recruitment, and disaster response, the report details ways in which responsible uses of technology can assist governments, businesses, NGOs, and migrant workers in preventing and mitigating the effects of labor trafficking (Latonero et al. 2015).

Another way to fight trafficking is to increase quantitative data and analysis. Human trafficking thrives in environments without data. Complex supply chains allow forced labor to remain hidden. If data collection and analysis are increased by GIS and geospatial technologies, coupled with mobile and network technologies, causes and trends can be examined so that support can be mobilized and action can be taken. With increased investigation, data collection, and data sharing, we can know about the specifics in which this transnational crime operates. Quantifying data also signifies the importance of a problem. In other words, what can be counted counts. Numbers can raise awareness and call attention to a hidden crime (Latonero et al. 2015). Some examples of emerging topics in gender inequalities related to labor trafficking include how technologies can be developed in ways that account for gender differences in labor; how the monetary flow of international remittances could potentially identify risk and exploitation; how humanitarian technologies used in crisis and disaster can be leveraged for human trafficking issues; what the useful technologies for migrants in conflict areas are; and what the ethical issues around data collection on migrants and vulnerable populations are.

GIS and the mapping process itself, especially as it relates to communities' identities, can be a challenging process as divergent gender identities exist. When gender identities overlap with other identities, they combine and intersect to generate distinct prejudices and discriminatory practices that violate individuals' equal rights in society. Intersectionality is the complex, cumulative way the effects of different forms of discrimination combine, overlap, or intersect – and are amplified when put together (IWDA 2018; Biernat and Sesko 2013; Purdie-Vaughns and Eibach 2008). A sociological term, "intersectionality" refers to the interconnected nature of social categories such as race, class, gender, age, ethnicity, ability, and residence status, regarded as creating overlapping and interdependent systems of discrimination or disadvantage. It emerges from the literature on civil legal rights. It recognizes that policies can exclude people. For example, MAP-Movement Advancement Project's Equality Maps survey the landscape of laws affecting LGBTQ individuals in the United States (Map 2020). Mapping LGBTQ equality sets out to identify and explain key gaps in gender and legal equality by introducing the major state and local laws and policies that protect or harm LGBT people. Each of the nearly 40 LGBTQ-related laws and policies that MAP tracks earn a score (positive for protective and negative for harmful). For each state, these individual policy scores are then added up to produce a summary tally score. Figure 1.3 shows the overall policy tallies (as distinct from sexual orientation or gender identity tallies) for each state, the District of Columbia, and the five populated U.S. territories. A state's "policy tally" counts the number of laws and policies within the state that help drive equality for

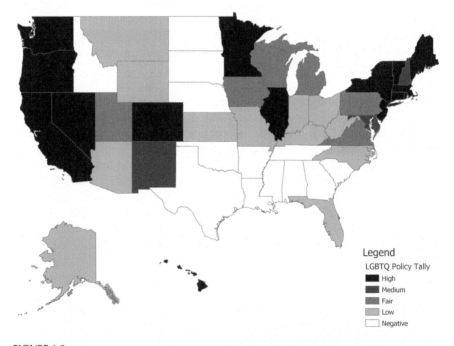

FIGURE 1.3
LGBTQ Policy Tally (Mapped by Spatial Analysis and Geographic Education Laboratory, University of Memphis, Data source: MAP-Movement Advancement Project's Equality Maps. (2020) https://www.lgbtmap.org/equality-maps/non_discrimination_laws. Accessed June 10, 2020).

LGBTQ people. The major categories of laws covered by the policy tally include: Relationship & Parental Recognition, Nondiscrimination, Religious Exemptions, LGBTQ Youth, Health Care, Criminal Justice, and Identity Documents.

MAP-Movement Advancement Project's Equality Maps reveal key LGBTQ issues such as nondiscrimination laws, education, employment, and health care. MAP provides data to the supporters of LGBTQ equality so that they can have a better understanding of the capacity and operations of the LGBTQ movement (MAP 2020). MAP's ongoing analysis of the movement's capacity provides organizations and their funders and supporters with information they can use to set goals and develop strategies, apply resources more effectively, and measure the success of the movement's work. Figure 1.4 shows the states with anti-bullying, nondiscrimination, and anti-LGBTQ laws.

As GIS helps communicate spatial trends, unfortunately, no global and updated data on the number of transgender persons are available, but there are approximately 1 million transgender adults living in the United States (Rawson and Williams 2014; Flores et al. 2016), and approximately 9.0 million living in Asia and the Pacific regions (Winter 2012; HPP 2015). Estimates from different world regions indicate that the prevalence of transgender identity varies between 0.1% and 1.1% of reproductive-age adults (UNAIDS 2014). Sweileh

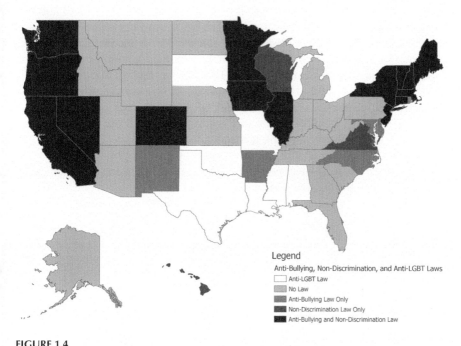

FIGURE 1.4
States with anti-bullying, nondiscrimination, and anti-LGBT laws (Mapped by Spatial Analysis and Geographic Education Laboratory, University of Memphis, Data source: MAP-Movement Advancement Project's Equality Maps. (2020) https://www.lgbtmap.org/equality-maps/no n_discrimination_laws. Accessed June 10, 2020).

FIGURE 1.5
Geographical distribution of peer-reviewed documents in transgender health (1900–2017). (Courtesy of, Sweileh, *BMC Int Health Hum Rights* 18, 16, 2018. Map source: the Faculty of Engineering at An-Najah National University. https://bmcinthealthhumrights.biomedcentral. com/articles/10.1186/s12914-018-0155-5/figures/3. Accessed June 10, 2020)

compiled a map of the geographical distribution of peer-reviewed documents in transgender health (1900–2017) (Sweileh 2017). The geographical distribution of the retrieved documents was based on the country affiliation of all authors participating in publishing the retrieved articles. Authors from 80 different countries contributed to publishing the retrieved documents. The retrieved documents originated mainly from Northern America, certain European countries, Australia, and Brazil (Figure 1.5). The African region, Middle East, and Eastern Europe had limited contribution to literature in transgender health.

GIS mapping seeks to express not only quantitative data but also qualitative data, such as shared attitudes or the connections among human experiences and places. Story-mapping technologies such as ArcGIS Story Map are a more-than-visual form of research representation (Wright 2014).

A story map visually displays data in relation to places, locations, or geographies. Story mapping helps to analyze complex social issues such as human rights, climate change, refugee resettlement, student transience, and community integration. Figure 1.6 is an example story map of the Sustainable Development Goals that guide audience through mapping with text, images, web links, video, and audio (UN-DESA 2019).

The Global Problem in Gender Inequality

This section covers a detailed description of gender inequality from a global perspective, examining how it still exists globally in spite of substantial and international measures that have been taken toward gender equality. What

FIGURE 1.6
A screenshot of an online story map on Sustainable Development Report 2019 (Source: Sustainable development report 2019 Story Map, ArcGIS Online, https://undesa.maps.arcgi s.com/apps/MapSeries/index.html?appid=48248a6f94604ab98f6ad29fa182efbd. Accessed June 9, 2020).

are the evident progresses? And what are the alarming issues regarding gender inequalities that still prevail today? And what makes gender inequality a global priority as a fundamental step in both human development and economic progress? This part of the chapter illustrates global and national level maps of measures of gender inequalities, such as access to basic education, health and life expectancy, equality of economic opportunity, and political empowerment.

The world's population faces immense challenges to sustainable development. Global health threats, global pandemics, climate change, more frequent and intense natural disasters, natural resource depletion, spiraling conflict, violent extremism, terrorism and related humanitarian crises, forced displacement of people, and unemployment threaten to reverse much of the development progress made in recent decades. The survival of many societies and the biological support systems of the planet are at risk. This leads to rising inequalities within and among countries. There are enormous disparities of opportunity, wealth, and power (UN General Assembly 2015). Gender inequality remains a key challenge and a major barrier to human development. The United Nations is now focusing its global development work on the recently developed 17 Sustainable Development Goals (UN 2015), which are listed in Table 1.1. The Goals demonstrate the scale and ambition of the universal 2030 Agenda that seeks to build on the Millennium Development Goals (UN General Assembly 2015) and complete what they did not achieve: to reach to all countries, all people, and all segments of society, while promising to prioritize the rights and needs of the most disadvantaged groups, from access to education and health care to clean water and decent work.

TABLE 1.1

UN Sustainable Development Goals 2019

Goal 1	End poverty in all its forms everywhere
Goal 2	End hunger, achieve food security and improved nutrition and promote sustainable agriculture
Goal 3	Ensure healthy lives and promote well-being for all at all ages
Goal 4	Ensure inclusive and equitable quality education and promote lifelong learning opportunities for all
Goal 5	Achieve gender equality and empower all women and girls
Goal 6	Ensure availability and sustainable management of water and sanitation for all
Goal 7	Ensure access to affordable, reliable, sustainable, and modern energy for all
Goal 8	Promote sustained, inclusive, and sustainable economic growth, full and productive employment and decent work for all
Goal 9	Build resilient infrastructure, promote inclusive and sustainable industrialization and foster innovation
Goal 10	Reduce inequality within and among countries
Goal 11	Make cities and human settlements inclusive, safe, resilient, and sustainable
Goal 12	Ensure sustainable consumption and production patterns
Goal 13	Take urgent action to combat climate change and its impacts
Goal 14	Conserve and sustainably use the oceans, seas, and marine resources for sustainable development
Goal 15	Protect, restore, and promote sustainable use of terrestrial ecosystems, sustainably manage forests, combat desertification, and halt and reverse land degradation and halt biodiversity loss
Goal 16	Promote peaceful and inclusive societies for sustainable development, provide access to justice for all and build effective, accountable, and inclusive institutions at all levels
Goal 17	Strengthen the means of implementation and revitalize the Global Partnership for Sustainable Development

Women have a critical role to play in all of the SDGs, with many targets specifically recognizing women's equality and empowerment as both the objective and a part of the solution. Goal 5, to "Achieve gender equality and empower all women and girls," is known as the stand-alone gender goal, because it is dedicated to achieving these ends. Girls and women have made major strides since 1990, but they have not yet gained gender equity. Access to education has greatly increased for both boys and girls, but the progress has been uneven, particularly in Africa, less developed countries, landlocked developing countries, and small island developing states. Some of the Millennium Development Goals remain off-track, in particular those related to maternal, newborn, and child health and reproductive health (UN General Assembly 2015). These disadvantages facing women and girls are a major source of inequality. All too often, women and girls are discriminated against in health, education, political representation, labor market, etc., with negative consequences for development of their capabilities and their freedom of choice. Deep legal and legislative changes are needed to ensure women's and girls' rights around the world. While a record 143 countries guaranteed equality between men and women in their constitutions by 2014, another 52 had not taken this step (UNDP 2019). Therefore, gender indicator data regularly produced by countries are critical to reach the Sustainable Development Goals targeting women's equality and empowerment. The list of the minimum set of gender indicators has been revised to be fully aligned with the sustainable development indicators. These qualitative and quantitative indicators address relevant issues to gender equality and/or women's empowerment. These indicators are classified into three tiers. Tier 1 is conceptually clear, with an internationally established methodology and standards, and data regularly produced by countries. Tier 2 is conceptually clear, with an internationally established methodology and standards, but data not regularly produced by countries. And Tier 3 has no internationally established methodology or standards, with data not regularly produced. Table 1.2 shows the minimum set of gender indicators (UNSC 2019).

The gender inequality index (GII) measures gender inequalities in three important aspects of human development: reproductive health, measured by maternal mortality ratio and adolescent birth rates; empowerment, measured by the proportion of parliamentary seats occupied by females and the proportion of adult females and males aged 25 years and older with at least some secondary education; and economic status, expressed as labor market participation and measured by the labor force participation rate of female and male populations aged 15 years and older (UNDP 2019). Figure 1.7 shows the structure of the gender inequality index to show how the GII is calculated. The index is a means to quantify the loss of achievement for a country due to gender inequality and is measured using three major factors: health, empowerment, and labor.

The GII is built on the same framework as the IHDI (Inequality-Adjusted Human Development Index) – to better expose differences in the distribution of achievements between women and men (UNDP 2019). The Gender inequality index was first introduced in the 2010 edition of the Human Development Report by the United Nations Development Programme (UNDP 2000). It measures the human development costs of gender inequality. Thus, the higher the GII value, the more the disparities between females and males and the more the loss to human development. It is a measurement of the difference between men and women in the respective country. Figure 1.8 sheds new light on the position of women in 162 countries; it yields insights into gender gaps in major areas of human development. The component indicators highlight areas in need of critical policy intervention, and they stimulate proactive thinking and public policy to overcome systematic disadvantages of women. The gender inequality index ranges from 0, where women and men fare equally, to 1, where one gender fares as poorly as possible in all measured dimensions. The countries with lower GII values have a higher success rate considering gender equality, as the GII measures the loss of success due to gender inequality. The worst-performing countries

TABLE 1.2

Gender Inequality Indicators

Indicator #	Indicator	Tier	Goal
1	Average number of hours spent on unpaid domestic and care work, by sex, age, and location	2	Goal 5
2	Average number of hours spent on total work (total work burden), by sex	2	Goal 5
3	Labor force participation rate for persons aged 15–24 and 15+, by sex	1	Goal 8
4	Proportion of employed who are own account workers, by sex	1	Goal 8
5	Proportion of employed who are contributing family workers, by sex	1	Goal 8
6	Proportion of employed who are employer, by sex	1	Goal 8
7	Percentage of adult population who are entrepreneurs, by sex	3	Goals 5 and 8
8	Percentage distribution of employed population by sector, each sex	1	Goal 8
9	Proportion of informal employment in non-agriculture employment, by sex	2	Goal 8
10	Unemployment rate, by sex, age, and persons with disabilities	1	Goal 8
11	Proportion of adults (age 15+) with an account at a bank or other financial institution or with a mobile money service provider, by sex	1	Goal 8
12	(a) Proportion of total agricultural population with ownership or secure rights over agricultural land, by sex, and (b) share of women	2	Goal 5
13	Gender gap in wages, by occupation, age, and persons with disabilities	2	Goal 8
14	Proportion of employed working part-time, by sex	2	Goal 8
15	Employment rate of persons aged 25–49 with a child age 3 living in a household and with no children living in the household, by sex	3	Goal 8
16	Proportion of children under age 3 in formal care	3	Goal 5
17	Proportion of individuals using the internet, by sex	1	Goal 17
18	Proportion of individuals who own a mobile telephone, by sex	2	Goal 5
19	Proportion of households with access to mass media, by sex of household head	3	
20	Youth literacy rate of persons (15–24 years), by sex	1	Goal 4
21	Adjusted net enrollment rate in primary education, by sex	1	Goal 4
22	Gross enrollment ratio in secondary education, by sex	1	Goal 4
23	Gross enrollment ratio in tertiary education, by sex	1	Goal 4
24	Gender parity index of the gross enrollment ratios in primary, secondary, and tertiary education	1	Goal 4

(Continued)

TABLE 1.2 (CONTINUED)

Gender Inequality Indicators

Indicator #	Indicator	Tier	Goal
25	Share of female science, technology, engineering, and mathematics graduates at tertiary level	1	Goal 4
26	Proportion of females among tertiary education teachers or professors	1	Goal 4
27	Adjusted net intake rate to the first grade of primary education, by sex	1	Goal 4
28	Primary education completion rate (proxy), by sex	1	Goal 4
29	Gross graduation ratio from lower secondary education, by sex	1	Goal 4
30	Effective transition rate from primary to secondary education (general programs), by sex	1	Goal 4
31	Educational attainment of the population aged 25 and older, by sex	1	Goal 4
32	Proportion of women of reproductive age (aged 15–49 years) who have their need for family planning satisfied by modern methods	1	Goal 3
33	Under-five mortality rate, by sex	1	Goal 3
34	Maternal mortality ratio	1	Goal 3
35	Antenatal care coverage	1	Goal 3
36	Proportion of births attended by skilled health professional	1	Goal 3
37	Age-standardized prevalence of current tobacco use among persons aged 15 and older, by sex	1	Goal 3
38	Proportion of adults who are obese, by sex	1	Goal 3
39	Number of new HIV infections per 1,000 uninfected population, by sex, age, and key populations	1	Goal 3
40	Access to anti-retroviral drug, by sex	1	Goal 3
41	Life expectancy at age 60, by sex	1	Goal 3
42	Mortality rate attributed to cardiovascular disease, cancer, diabetes, or chronic respiratory disease, by sex	1	Goal 3
43	Women's share of government ministerial positions	1	Goal 5
44	Proportion of seats held by women in (a) national parliaments and (b) local governments	1 (a) 2 (b)	Goal 5
45	Proportion of women in managerial positions	1	Goal 5
46	Percentage of female police officers	2	
47	Percentage of female judges	2	Goal 5
48	Proportion of ever-partnered women and girls aged 15 and older subjected to physical, sexual, or psychological violence by a current or former intimate partner in the previous 12 months, by form of violence and by age	2	Goal 5
49	Proportion of women and girls aged 15 and older subjected to sexual violence by persons other than an intimate partner in the previous 12 months, by age and place of occurrence	2	Goal 5

(Continued)

TABLE 1.2 (CONTINUED)

Gender Inequality Indicators

Indicator #	Indicator	Tier	Goal
50	Proportion of girls and women aged 15–49 years who have undergone female genital mutilation/cutting, by age	2	Goal 5
51	Proportion of women aged 20–24 years who were married or in a union before age 15 and before age 18	2	Goal 5
52	Adolescent birth rate (aged 10–14 years, aged 15–19 years) per 1,000 women in that age group	1	Goal 3

considering gender inequality are Yemen, the Central African countries, and Papa New Guinea. This is mostly due to the poor equality in empowerment and labor in these countries.

The Female Reproductive Health Index is used to assess a country's sexual and reproductive health and rights (Barot et al. 2015). This index consists of the maternal mortality rate and the adolescent birth rate and is assessed on the following four grounds: preventing unintended pregnancies; abortion care; help through pregnancy phases, and preventing and treating STI's. Figure 1.9 shows the care and support toward pregnant females, with lighter countries showing more support. Some countries have no data. As can be seen in Figure 1.9, Japan has a higher value than other countries in Asia because of the overall effectiveness of its health system and paralleled advances in technology in the country (WHO 2018a). In a similar way, Libya has a higher index value than its surrounding countries. This is mainly due to the history of Libya's medical education. In the early 1960s the foundation for its present relatively good health care system was laid by sponsoring medical students with scholarships ('Benamer and Bakoush 2009). Switzerland has the highest reproductive health index in the world (Expatica 2020). This is largely due to its mandatory health care insurance, which resulted in big investments in its health care system.

Maternal mortality rate is an essential indicator for improving maternal health. Death of a woman is classified as a maternal death when it occurs while she is pregnant or within 42 days of termination of pregnancy from any cause connected to or caused by the pregnancy, but not from accidental or incidental causes. It is classified as pregnancies resulting in deaths per 100,000 live births. Maternal deaths are easier classified this way when the cause of death attribution is not sufficient. Women below the age of 18 tend be at a higher risk for maternal mortality than woman aged between 18 and 19. Since 1990, maternal mortality has declined by 44%. Still, some 830 women and adolescent girls die each day from preventable causes related to pregnancy and childbirth. Figure 1.10 shows the mortality rate. According to the World Health Organization, 99% of all maternal deaths occur in developing countries, with more than half in fragile humanitarian settings (Klasen 2006). Additionally, maternal mortality is higher with women living in rural

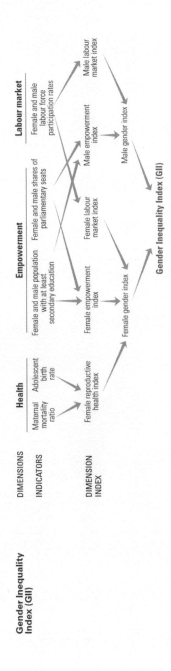

FIGURE 1.7

Gender Inequality Index: Dimensions and indicators (Source: From UNDP-United Nations Development Program, Human Development Report 2019, http://hdr.undp.org/en/content/gender-inequality-index-gii).

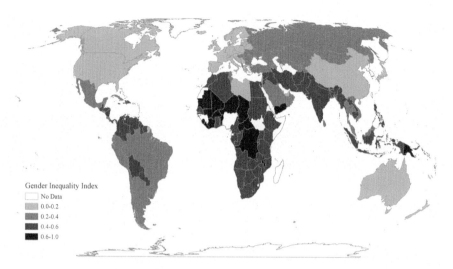

FIGURE 1.8
Map of gender inequality index (Mapped by Spatial Analysis and Geographic Education Laboratory, University of Memphis, Data source: UNDP-United Nations Development Program, Human Development Data 2019, http://hdr.undp.org/en/data. Accessed April 12, 2020).

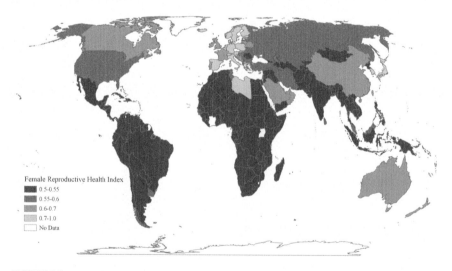

FIGURE 1.9
Female Reproductive Health Index (Mapped by Spatial Analysis and Geographic Education Laboratory, University of Memphis, Data source: UNDP-United Nations Development Program, Human Development Data 2019, http://hdr.undp.org/en/data).

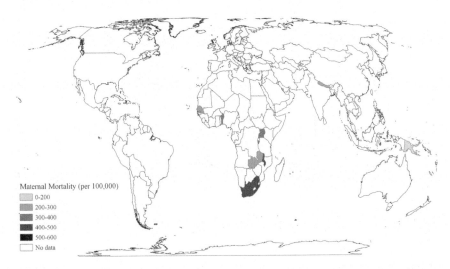

FIGURE 1.10
Maternal Mortality (Mapped by Spatial Analysis and Geographic Education Laboratory, University of Memphis, Data source: UNDP-United Nations Development Program, Human Development Data 2019, http://hdr.undp.org/en/data).

areas and among poorer communities. Both factors are the reason why Africa has a lot of maternal deaths. Haiti (521) in the western hemisphere and Sierra Leone (1165), South Africa (475), Malawi (439), Benin (391), and Uganda (336) in Africa have the highest deaths per 100,000 live births.

Adolescent birth rate, also referred to as age-specific fertility rate, is another essential indicator for improving maternal health. It is the number of annual live births to adolescent women per 1,000 adolescent women, with the age ranging between 15 and 19 years (WHO 2018b). Not only does early mother-hood affect the health of the mother, but it also undermines the opportunity for her to participate in proper education. A limitation in this indicator is that it is based on civil registration data, which is prone to inaccuracies or missing data, especially in the poorer regions. The countries in Africa and South America have on average a higher number of adolescent births in rela-tion to other countries. More specifically, countries in sub-Saharan Africa bear the highest risks of unintended pregnancy. There, the aforementioned arguments of education, investments, and other parallel technology stand as well. Mali, Niger, and Chad make up the top three countries on earth with the highest adolescent birth rate: 169.1, 192.0, and 161.1, respectively. Figure 1.11 shows adolescent females who gave birth per 1,000 women – the darker the color of the country the more the adolescent birth rate.

Many millions of girls face the prospect of child marriage and approxi-mately one in four girls in the developing world is married before the age of 18 (UNDP 2019). Figure 1.12 shows the median age at first marriage. The following are the top five countries with median age at first marriage below

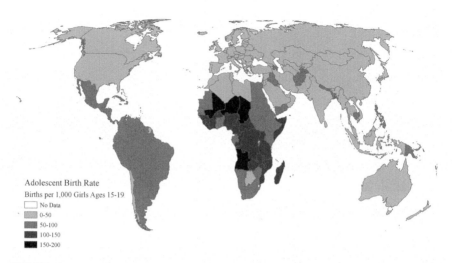

FIGURE 1.11

Adolescent birth rate (Mapped by Spatial Analysis and Geographic Education Laboratory, University of Memphis, Data source: UNDP-United Nations Development Program, Human Development Data 2019, http://hdr.undp.org/en/data).

18 years: Niger (15.7 years), Bangladesh (15.8 years), Yemen (16.0 years), Chad (16.1 years), and Ethiopia (16.1 years).

Family planning is central to women's empowerment and sustainable development. Unmet need for family planning is defined as the percentage of women of reproductive age, either married or in a union, who have an

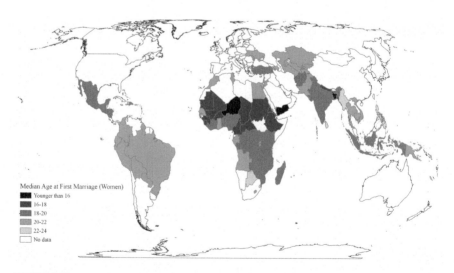

FIGURE 1.12

Median Age at first marriage (Mapped by Spatial Analysis and Geographic Education Laboratory, University of Memphis, Data source: UNDP-United Nations Development Program, Human Development Data 2019, http://hdr.undp.org/en/data).

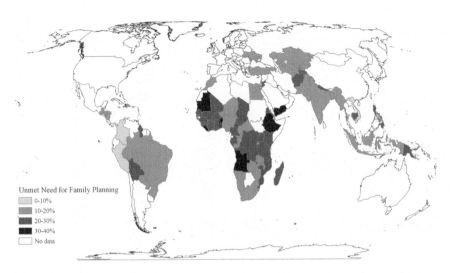

FIGURE 1.13
Unmet need for family planning (Mapped by Spatial Analysis and Geographic Education Laboratory, University of Memphis, Data source: UNDP-United Nations Development Program, Human Development Data 2019, http://hdr.undp.org/en/data. Accessed April 12, 2020).

unmet need for family planning. Women with unmet need are those who want to stop or delay childbearing but are not using any method of contraception. (Sonfield et al. 2014). Figure 1.13 shows the countries that have the highest unmet need for family planning: Yemen (40.0%); Angola (38.0%); Haiti (38.0%); Sao Tome and Principe (37.6%); and Ethiopia (36.1%) are the highest in the world.

The concept of unmet need points to the gap between women's reproductive intentions and their contraceptive behavior. Democratic and Health Surveys in 52 countries between 2005 and 2014 reveal the most common reasons that married women cite for not using contraception (Sedgh et al. 2016). Twenty-six percent of these women cite concerns about contraceptive side effects and health risks; 24% say that they have sex infrequently or not at all; 23% say that they or others close to them oppose contraception; and 20% report that they are breastfeeding and/or haven't resumed menstruation after a birth (Sedgh et al. 2016). Figure 1.14 shows that the countries where more than 10% of women cite any of these reasons are located in Western and Middle Africa, South America, Southeast Asia, and the Middle East. The countries that have the lowest percentage of married women currently using any method of contraception are Guinea (5.6%), Chad (5.7%), Eritrea (8.0%), Mauritania (8.0%), and Sierra Leone (8.2%).

Today, more than 300 million women in developing countries are using contraception, but more than 214 million women who want to plan their births do not have access to modern family planning (Alyahya 2019). Figure 1.15

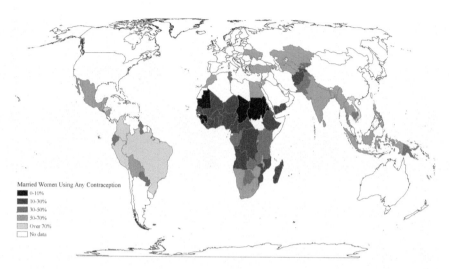

FIGURE 1.14
Married women using any contraception (Mapped by Spatial Analysis and Geographic Education Laboratory, University of Memphis, Data source: UNDP-United Nations Development Program, Human Development Data 2019, http://hdr.undp.org/en/data. Accessed April 12, 2020).

shows that Western and Middle African countries and the Middle East have the least demand for family planning satisfied by modern methods.

Sexual and reproductive health problems are a leading cause of death and disability for women in the developing world. Many millions of girls face harmful practices, such as female genital mutilation (UNICEF 2016). Figure 1.16

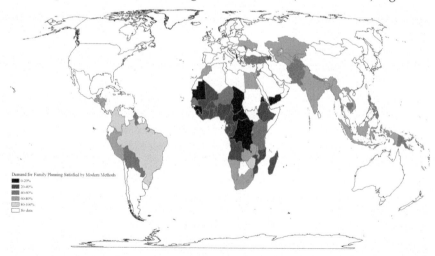

FIGURE 1.15
Demand for Family planning satisfied by modern methods (Mapped by Spatial Analysis and Geographic Education Laboratory, University of Memphis, Data source: UNDP-United Nations Development Program, Human Development Data 2019, http://hdr.undp.org/en/data).

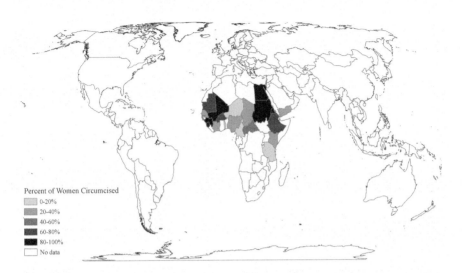

FIGURE 1.16
The percentage of women who are circumcised (Mapped by Spatial Analysis and Geographic Education Laboratory, University of Memphis, Data source: UNDP-United Nations Development Program, Human Development Data 2019, http://hdr.undp.org/en/data).

shows the percentage of women circumcised, with the top five countries being in Africa: Guinea (96.9%), Eritrea (94.5%), Egypt (92.3%), Mali (91.4%), and Sierra Leone (91.3%).

HIV disproportionately affects women and adolescent girls because of the vulnerabilities created by unequal cultural, social, and economic status. Much has been done to reduce mother-to-child transmission of HIV, but much more needs to be done to reduce the gender inequality and violence that women and girls at risk of HIV often face. Women account for more than half the number of people living with HIV worldwide. Young women (10–24 years old) are twice as likely to acquire HIV as young men the same age (UNAIDS 2019). Sub-Saharan African countries bear the highest risks of HIV infection among women (Ramjee, and Daniels 2013). Figure 1.17 shows HIV prevalence in African counties, with Swaziland (31.1%), Lesotho (29.7%), South Africa (27.3%), Namibia (16.9%), and Zimbabwe (16.7%) representing the top five.

Unaccommodating attitudes toward sex outside of marriage and the restricted social autonomy of women and young girls reduce their ability to access sexual health and HIV services (UNAIDS 2019). Figure 1.18 shows that the countries with the highest percentages of women receiving HIV tests results in the past 12 months are the following: Zambia (64.1%), South Africa (58.5%), Lesotho (58%), Uganda (54.6%), and Namibia (49.1%).

As the struggle for gender equality continues, violence against women and girls remains a global crisis affecting all countries. One in three women will experience physical or sexual violence in her lifetime. According to WHO, violence is a pervasive public health and human rights problem around the

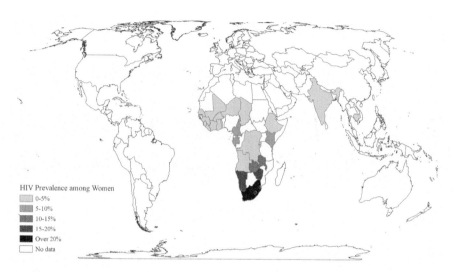

FIGURE 1.17
HIV prevalence among women (Mapped by Spatial Analysis and Geographic Education Laboratory, University of Memphis, Data source: UNDP-United Nations Development Program, Human Development Data 2019, http://hdr.undp.org/en/data).

world. It affects women, men, boys, and girls in all countries and cuts across boundaries of age, race, religion, ethnicity, disability, culture, and wealth. Statistically, women and children (both boys and girls) are most affected by violence in the home and it is often perpetrated by men they know and trust (WHO 2013). Worldwide, 35% of women have experienced physical and/or

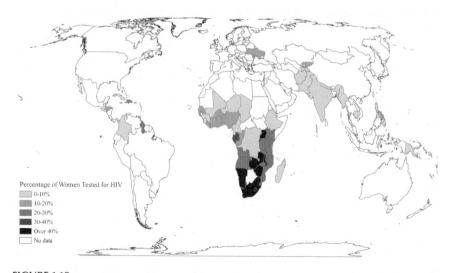

FIGURE 1.18
The percentage of women tested for HIV (Data source: UNDP-United Nations Development Program, Human Development Data 2019, http://hdr.undp.org/en/data).

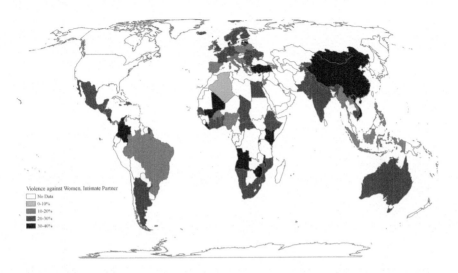

FIGURE 1.19
Intimate partner violence against women (Mapped by Spatial Analysis and Geographic Education Laboratory, University of Memphis, Data source: UNDP-United Nations Development Program, Human Development Data 2019, http://hdr.undp.org/en/data).

sexual intimate partner violence or non-partner sexual violence. Figure 1.19 shows that intimate partner violence exists even in countries that have made laudable progress in other areas, such as European and Scandinavian countries, China, and Australia. Unfortunately, comparable cross-national data on these forms of violence are not available in the United States and Canada because of reporting practices and the lack of intra-country data specifically in the context of intimate partner violence (CDC 2014; Statistics Canada 2016).

The UN system continues to give attention to the issue of violence against women. Women and girls are disproportionately subjected to violence, including femicide, sexual violence, intimate partner violence, trafficking, and harmful practices. In September 2017, the European Union and the United Nations joined forces to launch the Spotlight Initiative, a global, multi-year initiative that focuses on eliminating all forms of violence against women and girls. The Initiative is so named as it brings focused attention to this issue, moving it into the spotlight and placing it at the center of the efforts to achieve gender equality and women's empowerment, in line with the 2030 Agenda for Sustainable Development (Spotlight Initiative 2018). Figure 1.20 suggests violence against women in Mexico and the Middle and South Americas take many forms and is substantial. The WHO estimates that more than 1 in 10 women aged 15 years and older in Latin American counties have experienced forced sex by a non-partner (Mendoza et al. 2014). Other forms of sexual violence include sexual harassment in workplaces and public spaces, noncontact sexual abuse and exploitation, trafficking, and sexual violence in the context of armed conflict, as documented in Colombia (UN 2016).

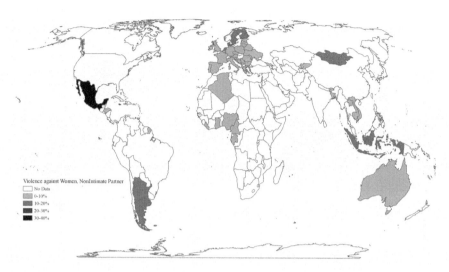

FIGURE 1.20
Non-intimate partner violence against women (Mapped by Spatial Analysis and Geographic Education Laboratory, University of Memphis, Data source: UNDP-United Nations Development Program, Human Development Data 2019, http://hdr.undp.org/en/data).

Stark gender disparities remain in economic and political realms. The empowerment index seeks to measure relative female representation in economic and political power. It considers gender gaps in education, representation in political and professional positions, and gender gaps in incomes (Raj 2017). The first indicator of empowerment is the population with at least secondary education. Figure 1.21 shows the percentage of women who have

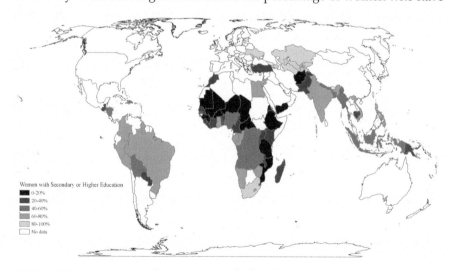

FIGURE 1.21
Women with secondary or higher education (Data source: UNDP-United Nations Development Program, Human Development Data 2019, http://hdr.undp.org/en/data).

at least a secondary-level education. In correlation with the gender inequality index, countries that do not support educating their female population have a lower GII value. The countries with the lowest percentage of women with secondary education are Yemen (7.1%), Niger (8.5%), Afghanistan (8.6%), Ethiopia (11.9%), and Burkina Faso (12.4%). In contrast, there are several countries with a 100% education rate for both the female and male population: Austria, Iceland, Canada, Finland, Estonia, and Luxembourg.

Education is vital for the literacy of a country's population. The global literacy rate for all people aged 15 years and above is 86.3%. The global literacy rate for all males is 90.0% and the rate for all females is 82.7% (UIS 2019). Figure 1.22 shows that the countries with the lowest literacy rate among women are Niger (14.0%), Afghanistan (14.8%), Mali (20.6%), Chad (22.1%), and Burkina Faso (22.5%).

Many countries nowadays tend to have significant seats occupied by female politicians to support the equal importance of women when it comes to making decisions for the country. As of 2018, only 24% of all national parliamentarians were female, a slow rise from 11.3% in 1995 (IPU and UN Women 2019). For most of history, the governments were predominantly male, with little to no place for women to voice their opinion. With the increase in gender equality in the workplace, there has also been an increase in gender equality in the government. Figure 1.23 shows the number of seats occupied by females in the government. Rwanda is one of the only countries on earth that has more women than men in the parliament: 55.70%. This is because it is required by law to have equal seating for the two genders, but the women in the parliament and at home are still not being treated with the respect they deserve (UNWomen 2018). On the

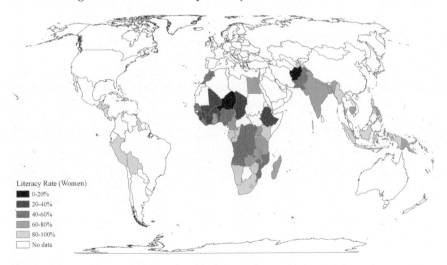

Literacy Rate (Women)
- 0-20%
- 20-40%
- 40-60%
- 60-80%
- 80-100%
- No data

FIGURE 1.22
Literacy rate among women (Mapped by Spatial Analysis and Geographic Education Laboratory, University of Memphis, Data source: UNDP-United Nations Development Program, Human Development Data 2019, http://hdr.undp.org/en/data).

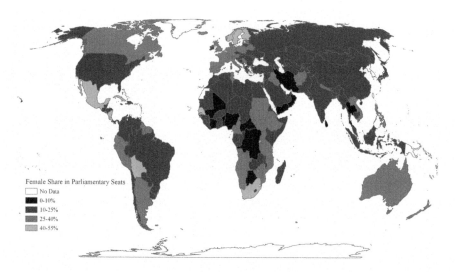

FIGURE 1.23
Female share in parliamentary seats (Mapped by Spatial Analysis and Geographic Education Laboratory, University of Memphis, Data source: UNDP-United Nations Development Program, Human Development Data 2019, http://hdr.undp.org/en/data).

other side, there are three countries with not a single woman in the parliament: Papua New Guinea, Vanuatu, and Micronesia. There are several reasons why women do not have any seats in the parliament in Papua New Guinea (IWDA 2017). First, many Papua New Guineans perceive politics as a man's world and find women ill-equipped to be effective political leaders. Secondly, the current political culture promotes corrupt practices. Thirdly, there are no adequate financial resources and logistical support for the entire election period. Finally, women struggle to find support of traditional tribal leaders to secure enough votes to be a serious contender to male candidates.

The goal of the Labor Market Rate index is to represent the spread of the genders in the workplace (UNDP 2019). This is an important index to show how many women are active on the labor market with respect to men in a certain country. While there has been some progress over the decades, on average, women in the labor market still earn 20% less than men globally. Figure 1.24 shows female labor force participation. An interesting thing can be seen in the south of Africa. Around the 1960s the participation rates in Angola, Mozambique, and Guinea-Bissau were quite low (7%, 7%, and 1%, respectively) (Durand 1975). This was the result of a statistical convention applied under Portuguese rule that excluded unpaid family work in the participation rate. Following independence from Portugal, the participation rates increased and the growth in labor participation in Mozambique was almost 10% per decade between 1960 and 1980 (Tzannatos 1999). Of the total female population in Mozambique, 82.50% are active on the labor market, in comparison to 74.60% of the male population.

Share in managerial positions is an additional indicator which provides information on the proportion of men and women who are employed in

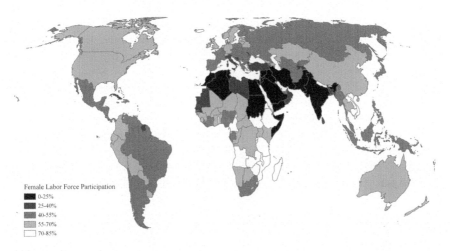

FIGURE 1.24
Female labor force participation (Mapped by Spatial Analysis and Geographic Education Laboratory, University of Memphis, Data source: UNDP-United Nations Development Program, Human Development Data 2019, http://hdr.undp.org/en/data).

decision-making and management roles in government, large enterprises, and institutions. This provides some insight into women's power in decision-making and the economy (Holst 2006). In a way, this indicator is a combination of the labor and empowerment dimension. Figure 1.25 shows the share in managerial positions – the lighter the color, the closer the countries are to a 50/50 ratio in managerial positions. Only a limited number of countries have

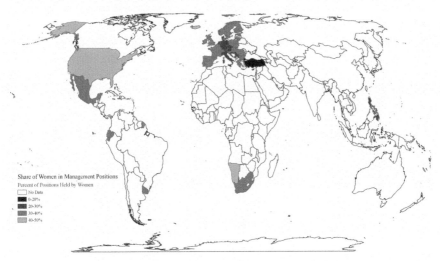

FIGURE 1.25
Share in managerial positions (Mapped by Spatial Analysis and Geographic Education Laboratory, University of Memphis, Data source: UNDP-United Nations Development Program, Human Development Data 2019, http://hdr.undp.org/en/data).

data about this indicator. Nevertheless, Europe is well represented. In Europe, 30% of managerial positions are occupied by women on average, but there is one remarkable country. Although the surrounding countries have an average score, Luxembourg has a relatively low representation of women regarding top managerial positions: 14.90%. To tackle this issue, the Luxembourg government presented a strategy for a better gender balance in management layers in September 2014 (Theis 2019). Mali has the lowest share of females in managerial positions because women are typically paid less than men and sometimes are not listed as employees at all (AFDB 2011). Countries around the equator seem to have a lower labor force participation rate for females. For the female share in managerial positions, it can be concluded that Africa performs the worst.

Conclusion: Gender Equality Cannot Wait!

It is a time of immense opportunity for gender equality in the world. GIS technology is the key to expanding our understanding of where we are in terms of gender equality, how we got here, and the path forward. The lowest gender inequality can be found in Europe, closely followed by North America. Africa does not score well in any gender equality indicator, with the maternal mortality rate being the relatively worst indicator. Oceania can improve a lot when it comes to empowerment, considering there are no females who occupy a seat in the parliament in, for example, Papua New Guinea. Asia has an overall mediocre performance. South America is the second worst-performing continent, mainly due to its low scores on the health indicators. Women are still disadvantaged in many countries, with Africa scoring the lowest for most indicators of the gender inequality index. All the countries, including the high-scoring countries in Europe, need to work to prevent and respond to gender-based violence with policymakers, justice systems, health systems, and humanitarian partners. Governments need to promote universal access to quality, integrated sexual and reproductive health services and strengthen health systems and reporting practices to respond to women subjected to intimate partner violence and/or sexual violence. There is still a lot of work to do to eliminate harmful practices, including female genital mutilation (FGM) and child marriage, and to engage men and boys to advance gender equality.

Gender equality is a milestone, and progress depends on inclusion. Inclusion begins with all countries, all people, and all segments of society, while promising to address the rights and needs of the most disadvantaged groups. As key gender indicators are collected and mapped by GIS at subnational levels, and finer spatial resolution of gender data and surveys become available, big data will lead to new avenues for detailed spatial investigations. Successful emerging technology solutions such as the cloud, social media, cell phones, and internet usage will require creative and transformative approaches to gender equality. GIS technologies can be leveraged for human trafficking issues, migrants in conflict areas, vulnerable populations, and better understanding of LGBT equality and the LGBT movement.

The world's milestones and performance metrics to reach the sustainable goals of the 2030 Agenda constitute task-driven intermediate steps toward gender equality. The goal is for all countries, all people, and all segments of society to be equal, for everyone to belong, and for everyone to be connected. If any of us suffers, we suffer together. Gender equality benefits the whole world, men and women, no matter their level of education, privilege, or accomplishments, in the home or the workplace. There is no wrong race or religion or gender. No one is exploited because they are poor or excluded because they are weak.

GIS has great potential to accelerate human progress, to bridge the digital divide, and to develop knowledge societies to reach gender equality. In this book, we put gender on the map from Mexico to Nigeria and to China and present issues from gender inequity in the workplace to intimate partner violence. In our globally interconnected world, connectedness means conversation, a great deal of conversation. We become each other's databases, servers, and maps, leaning on each other's memories, multiplying, amplifying, and anchoring the things we could imagine by sharing our dreams, our speculations, and our curiosities. With this book, we share the stories of disadvantaged people from different parts of the world who have given urgency toward gender equality and to change the world – and most importantly, ourselves.

References

AFDB-African Development Bank. (2011). *Gender in Employment: Case Study of Mali* 1(1). https://www.afdb.org/fileadmin/uploads/afdb/Documents/Publication s/Gender%20in%20Employment_Case%20Study%20of%20Mali%20_for%2 0distribution.pdf

Ahmed S., Hill K. (2011). Maternal mortality estimation at the subnational level: A model-based method with an application to Bangladesh. *Bulletin World Health Organization* 89(1): 12–21.

Alyahya M.S., Hijazi H.H., Alshraideh H.A. (2019). Do modern family planning methods impact women's quality of life? Jordanian women's perspective. *Health and Quality of Life Outcomes* 17: 154.

Barot S., Cohen S., Darroch J., Galati A.J., Polis C., Starss S., Starss A.M. (2015). *Female Reproductive Health Index. SRHR-Sexual and Reproductive Health and Rights Indicators for the SDGs: Recommendations for Inclusion in the Sustainable Development Goals and the Post-2015 Development Process*. Guttmacher Institute. http://www.guttmacher.org

Benamer, H.T.S., Bakoush, O. (2009). Medical education in Libya: the challenges. *Medical Teacher* 31(6): 493–496, doi:10.1080/01421590902832988

Bhutta Z.A., Reddy K. (2012). Achieving equity in global health: so near and yet so far. *JAMA* 307(19): 2035–2036.

Biernat M., Sesko A.K. (2013). Evaluating the contributions of members of mixed-sex work teams: race and gender matter. *Journal of Experimental Social Psychology* 49(3): 471–476.

Burgert C.R. (2014). *Spatial Interpolation with Demographic and Health Survey Data: Key Considerations*. Rockville: ICF International.

CDC. (2014). United States centers for disease control and prevention. In *Intimate Partner Violence in the United States – 2010*. Atlanta: National Center for Injury Prevention and Control, Division of Violence Prevention. http://www.cdc. gov/violenceprevention/pdf/cdc_nisvs_ipv_report_2013_v17_single_a.pdf

Cromley E.K. (2003). GIS and disease. *Annual Review of Public Health* 24(1): 7–24.

Detres M., Lucio R., Vitucci J. (2014). GIS as a community engagement tool: developing a plan to reduce infant mortality risk factors. *Maternal and Child Health Journal* 18(5): 1049–1055.

DHS-Demographic and Health Surveys Program. *Spatial Data Repository*, spatialdata. dhsprogram.com. Accessed April 14, 2020.

Digital Globe. (2020). https://www.digitalglobe.com/. Accessed May 10, 2020.

DIVA-GIS. (2020). *Spatial Data*. http://www.diva-gis.org/Data. Accessed June 5, 2020.

Donovan M. (2017). *Technology in the Fight Against Trafficking: Tracking Criminals and Helping Victims*. National Consumers League Report, Child Labor Coalition Program. Washington, DC. https://www.nclnet.org/trafficking_tech. Accessed January 17, 2020.

Durand A. (1975). *The Labor Force in Economic Development: A Comparison of International Census Data, 1946–1966*. Princeton, NJ: Princeton University Press.

Ebener S, Guerra-Arias M, Campbell J, Tatem A, Moran A, Amoako Johnson F, Fogstad H, Stenberg K, Neal S, Bailey P, Porter R, Matthews Z. (2015). The geography of maternal and newborn health: the state of the art. *International Journal of Health Geographics* 14(1): 19.

Expatica. 2020. *A Guide to Healthcare in Switzerland*. https://www.expatica.com/ch/he althcare/healthcare-basics/healthcare-in-switzerland-103130/. August 6, 2020.

Flores A.R., Herman J.L., Gates G.J., Brown T.N.T. (2016). *How Many Adults Identify as Transgender in the United States*. Williams Institute. University of UCLA. https ://williamsinstitute.law.ucla.edu/wp-content/uploads/Trans-Adults-US-A ug-2016.pdf

GDWGL-Global Donor Working Group on Land. (2020). https://www.donorplatform .org/land-governance.html. Accessed May 5, 2020.

GLGP –The Global Land Governance Programme Map & Database. (2020). https:// landgov.donorplatform.org/. Accessed May 5, 2020.

Gozdyra P, Glazier R, Moldofsky B. (1999). *A Picture Speaks a Thousand Numbers: Allowing the Community to Examine Available Health Data Through User Friendly Mapping Software*. New Zealand: University of Otago.

Holst E. (2006). Women in managerial positions in Europe: focus on Germany. *Management Review* 17(2): 122–142. JSTOR, http://www.jstor.org/stable/41771572. Accessed 8 June 2020.

HPP-Health Policy Project, Asia Pacific Transgender Network, United Nations Development Programme. (2015). *Blueprint for the Provision of Comprehensive Care for Trans People and Trans Communities*. Washington, DC: Futures Group, Health Policy Project.

https://www.land-links.org/project/mobile-application-to-secure-tenure-tanzania/. Accessed June 8, 2020.

IPU-Inter-Parliamentary Union and UN Women. (2019). *Women and Politics 2019 Map.* https://www.unwomen.org/en/digital-library/publications/2019/03/women-in-politics-2019-map

IWDA-International Women's Development Agency. (2017). *Why Were Zero Women Elected to Papaua New Guinea's Parliament?* https://iwda.org.au/where-were-the-women-in-papua-new-guineas-election/. Accessed October 17, 2017.

IWDA-International Women's Development Agency. (2018). https://iwda.org.au/assets/files/IWDA-2018-Annual-Report.pdf. Accessed May 14, 2020.

Klasen S. (2006) UNDP's gender-related measures: some conceptual problems and possible solutions. *Journal of Human Development* 7(2): 243–274.

Latonero M., Wex, B., Dank M. (2015). *Technology and Labor Trafficking in a Network Society : General Overview, Emerging Innovations and Philippines Case Study Report. USC Annenberg Center on Communication Leadership and Policy.* https://technologyandtrafficking.usc.edu/technology-labor-trafficking-network-society/. Accessed May 12, 2020.

MAP-Movement Advancement Project's Equality Maps. (2020) https://www.lgbtmap.org/equality-maps/non_discrimination_laws. Accessed June10, 2020.

Mendoza J.A., Bott S., Guedes A., Goodwin M. (2014). Intergenerational effects of violence against girls and women: Selected findings from a comparative analysis of population-based surveys from 12 countries in Latin America and the Caribbean. In Dubowitz H, editor. *World Perspectives on Child Abuse.* 10th ed. Aurora: International Society for Prevention of Child Abuse and Neglect, 124–133.

Molla Y.B., Rawlins B., Makanga P.T. (2017). Geographic information system for improving maternal and newborn health: Recommendations for policy and programs. *BMC Pregnancy Childbirth* 17(1): 26.

Murai S., Singh R.B. (1998). *Space Informatics for Sustainable Development.* Brookfield, VT: A.A. Balkema Publishers.

Purdie-Vaughns V., Eibach R.P. (2008). Intersectional invisibility: the distinctive advantages and disadvantages of multiple subordinate-group identities. *Sex Roles: A Journal of Research* 59(5–6): 377–391.

Raj A. (2017). Gender empowerment index; a choice of progress or perfection. *The Lancet Global Health* 5: 849–850.

Ramjee G., Daniels B. (2013). Women and HIV in Sub-Saharan Africa. *AIDS Research and Therapy* 10(1): 30.

Rawson K., Williams C. (2014). Transgender: the rhetorical landscape of a Term. *Present Tense* 3(2).

Ruktanonchai C., Pindolia D., Striley C., Odedina F., Cottler L. (2014). Utilizing spatial statistics to identify cancer hot spots: a surveillance strategy to inform community-engaged outreach efforts. *International Journal of Health Geographics* 13(1): 39.

Sedgh G., Ashford L.S., Hussain R. (2016). *Unmet Need for Contraception in Developing Countries: Examining Women's Reason's Not Using a Method.* New York: Guttmacher Institute, http://www.guttmacher.org/report/unmet-need-for-contraception-in-develing-countries. Accessed June 10, 2020.

Sonfield A., Hasstedt K., Gold R.B. (2014). *Moving Forward: Family Planning in the Era of Health Reform.* New York: Guttmacher Institute.

Spotlight Initiative. (2018). *Annual Report 2018*. European Union and the United Nations. http://www.spotlightinitiative.org/. Accessed June 10, 2020.

Statistics Canada. (2016). Family *Violence* in Canada: *A Statistical Profile*, 2014. http://www.statcan.gc.ca/pub/85-002-x/2016001/article/14303-eng.htm

Sweileh W.M. (2017). Bibliometric analysis of peer-reviewed literature in transgender health (1900 – 2017). *BMC International Health and Human Rights* 18: 16.

Theis F. (2019). *Apropos. Luxembourg and the European Union. Information and Press Service of the Luxembourg Government*. https://luxembourg.public.lu/en/publications/ap-europe.html. Accessed June 10, 2020.

UIS-UNESCO Institute for Statistics. (2019). *Adult Literacy Rate, Population 15+ Years Both Sexes, Female, Male*. 20 November. http://data.uis.unesco.org/index.aspx?queryid=166

UN. (2016). Report of the Secretary-General on *Conflict-Related Sexual Violence*. New York: United Nations Security Council. http://www.securitycouncilreport.org/atf/cf/%7B65BFCF9B-6D27-4E9C-8CD3-CF6E4FF96FF9%7D/s_2015_203.pdf

UNAIDS. (2014). *The Gap Report 2014 (Transgender People). Beginning of the End of the Aids Epidemic*. Geneva, Switzerland: Joint United Nations Programme on HIV/AIDS (UNAIDS).

UNAIDS. (2019). Women and HIV_A spotlight on adolescent girls and young women, p6. https://www.unaids.org/sites/default/files/media_asset/2019_women-and-hiv_en.pdf Accessed June 10, 2020.

UN-DESA United Nations Department of Economic and Social Affairs. (2019). *Sustainable Development Report 2019 Story Map, ArcGIS*. https://undesa.maps.arcgis.com/apps/MapSeries/index.html?appid=48248a6f94604ab98f6ad29fa182efbd. Accessed June 9, 2020.

UNDP. (2000). *Human Development Report 2000: Human Rights and Human Development*. http://www.hdr.undp.org/en/content/human-development-report-2000

UNDP-United Nations Development Program. (2019). *Human Development Report 2019. Beyond Income, Beyond Averages, Beyond Today: Inequalities in Human Development in the 21st Century*. New York. http://hdr.undp.org/sites/default/files/hdr2019.pdf

UNFPA ICM-United Nations Population Fund, The International Confederation of Midwives (2014). *The State of the World's Midwifery 2014: A Universal Pathway. A Women's Right to Health*. New York: United Nations Population Fund.

UNICEF (2016). Female Genital Mutilation/Cutting: A Global Concern UNICEF. New York. https://www.unicef.org/media/files/FGMC_2016_brochure_final_UNICEF_SPREAD.pdf

UNSC-United Nations Statistical Commission. (2019). *Tier Classification for Global SDG Indicators*. https://unstats.un.org/sdgs/files/Tier%20Classification%20of%20SDG%20Indicators_17%20April%202020_web.pdf. Accessed April 17, 2020.

UN-United Nations General Assembly. (2015). *Transforming Our World: the 2030 Agenda for Sustainable Development*. New York: United Nations Security Council; Sustainable development Goals Knowledge Platform. https://www.un.org/ga/search/view:doc.asp?symbol=A/RES/70/1&Lang=E

UN-United Nations Sustainable Development Summit. (2015). http://www.un.org/sustainabledevelopment/health/. Accessed June 5, 2020.

UN-United Nations-The Sustainable Development Goals Report. (2019). New York: United Nations. https://unstats.un.org/sdgs/report/2019/The-Sustainable -Development-Goals-Report-2019.pdf. Accessed June 2, 2020.

UNWomen (2018). *Revisiting Rwanda Five Years After Record-Breaking Parliamentary Elections.* https://www.unwomen.org/en/news/stories/2018/8/feature-rwand a-women-in-parliament. Accessed August 13, 2018.

USAID Land Projects. (2016). *Tanzania: Mobile Application to Secure Tenure.* Burlington, VT: United States Agency for International Development Office of Land and Urban.

WHO. (2013). *Global and Regional Estimates of Violence Against Women: Prevalence and Health Effects of Intimate Partner Violence and Non-Partner Sexual Violence.* Geneva. http:// www.who.int/reproductivehealth/publications/violence/9789241564625/en/.

WHO. (2018a). Regional office for South-East Asia. Japan health system review. *Health Systems in Transition* 8(1), 10–14.

WHO. (2018b). *Survive, Thrive, Transform. Global Strategy for Women's, Children's and Adolescents' Health: 2018 Report on Progress Towards 2030 Targets.* Geneva: World Health Organization.

Winter S. (2012). *Lost in Transition: Transgender People. Rights and HIV Vulnerability in the Asia-Pacific Region Focus, UNDP Asia Pacific Transgender Network.* 69. http: //www.weareaptn.org/wp-content/uploads/2017/10/lost-in-transition.pdf. Accessed June 8, 2020.

Wright D.J. (2014). Story Maps as an Effective Social Medium for Data Synthesis, Communication, and Dissemination. ArcGIS Online. doi:10.13140/ RG.2.2.10448.97283.

Zafiris Tzannatos. (1999). Women and labor market changes in the global economy. *Growth Helps, Inequalities Hurt and Public Policy Matters* 27(3), 551–569. https://do i.org/10.1016/S0305-750X(98)00156-9

Resource Guide of Organizations that Provide Gender Indicator Data

Organizations whose websites readers can visit and download gender indicators data are listed in this section.

UN Gender Statistics

https://genderstats.un.org/#/home

Collection of 52 quantity and 11 quality indicators addressing relevant issues related to gender equality and women's empowerment.

United Nations Development Program

The United Nations Development Programme is the United Nations' global development network. It advocates for change and connects countries to knowledge, experience, and resources to help people build a better life for themselves.
http://hdr.undp.org/en/data

UNdata

UNdata is an internet-based data service which brings UN statistical databases within easy reach of users through a single-entry point.
http://data.un.org/

UNWomen

The United Nations Entity for Gender Equality and the Empowerment of Women, also known as UN Women, is a United Nations entity working for the empowerment of women. UN Women became operational in January 2011.
https://www.unwomen.org/en/how-we-work/research-and-data

GHO – Global Health Observatory Data

The GHO data repository contains an extensive list of indicators, which can be selected by theme or through a multi-dimensional query functionality. The aim of the GHO portal is to provide easy access to country data and statistics with a focus on comparable estimates and that of WHO's is to monitor global, regional, and country-specific situations and trends.
https://www.who.int/data/gho

Family Planning 2020

FP2020 is a global partnership to empower women and girls by investing in rights-based family planning.
http://www.familyplanning2020.org/data-hub

Girls Not Brides

Girls Not Brides is a global partnership of 1000+ civil society organizations committed to ending child marriage and enabling girls to fulfill their potential.
https://www.girlsnotbrides.org/resource-centre/

#MeToo movement

The "me too" movement supports survivors of sexual violence and their allies.
The "me too" Healing Resource Library is a comprehensive database consisting of local and national organizations dedicated to providing services and safe spaces for survivors of sexual violence.
https://metoomvmt.org/resources/

Population Council

The Population Council conducts research to address critical health and development issues in more than 50 countries.
https://www.popcouncil.org/research

Save the Children

Save the children works worldwide to inspire breakthroughs in the way the world treats its children and to achieve immediate and lasting change in their lives.
https://www.savethechildren.org/us/what-we-do/events/bridge-the-gap-for-girls

MAP-Movement Advancement Project's Equality Maps.

Founded in 2006, the Movement Advancement Project (MAP) is an independent, nonprofit think tank that provides rigorous research, insights, and communications that help speed equality and opportunity for all.
https://www.lgbtmap.org/equality-maps

2

Mapping Domestic Violence

Amaia Iratzoqui, James C. McCutcheon, and Angela D. Madden

CONTENTS

Introduction

Historically, domestic violence has been treated as a unique form of violence, both within criminological research and in the field. In this chapter, the term "domestic violence" (DV), often used interchangeably with "intimate partner domestic violence" (IPDV), refers to physical, sexual, psychological, emotional, and other forms of violence between and by current and former romantic and marital partners. Criminologists long have known that DV is one of the most underreported crimes; estimates suggest that at least one-third of all DV cases are not reported to the police (Truman and Morgan 2016). Law enforcement traditionally has treated DV, particularly physical DV, as a form of violence tied to the home, rather than as one of the many types of violent behavior. Current research, however, has provided evidence of similarities between patterns of DV and other forms of violence (Iratzoqui and McCutcheon 2018).

Geographic information systems (GIS) technology is increasingly being adapted for criminal justice response. Geospatial analyses of crime have been used to identify "hot spots" of criminal activity, direct police patrol, and examine links between structural, cultural, and individual-level factors and criminal incidents (Levine 2006). More recently, research has called for the application of the GIS technology to DV, including recommendations to map community- and neighborhood-level influences on various measures of IPDV occurrences (Murray, Bunch, and Hunt 2016), as well as victim access to resources after victimization (Peek-Asa, Wallis, Harland, Beyer, Dickey, and Saftlas 2011).

One way to better understand the nature of DV, then, is to focus attention on how crime mapping can (1) improve law enforcement response, (2) increase victim reporting, and (3) increase victim access to services. These tasks are of utmost urgency in cities with large disadvantaged populations suffering economic deprivation, many of which are currently experiencing crime increases.

Economics plays a significant role in many of the conditions, including health, safety, and crime. Memphis, Tennessee, a city with many opportunity zones within its metropolitan area, has a poverty rate of 24.6%, a rate that continues to be relatively stagnant. Figure 2.1 shows a standard deviation map of the distribution of poverty in Memphis, ranging from 1.5 standard deviations below the mean to 2.5 standard deviations above the mean; areas that are significant are being compared in relation to the mean of the various census tracts in Memphis.

Figure 2.2 shows areas above and below the average poverty rate on a census tract level map; green areas are at or below the average poverty rate, while red areas are above the mean poverty rate for the city. Previous research that has included data from Memphis has linked economic disadvantage (i.e., poverty, unemployment, median household income) to violent crime (Burraston, McCutcheon, and Watts 2018). The current chapter builds upon this work by using GIS to map the locations of DV calls for service,

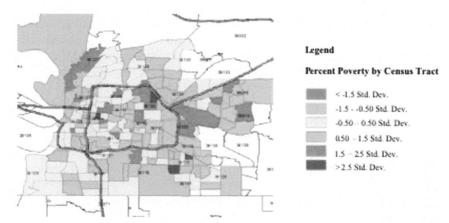

FIGURE 2.1
Standard deviation map of the distribution of poverty, 2015, in Memphis, ranging from 1.5 standard deviations below the mean to 2.5 standard deviations above the mean.

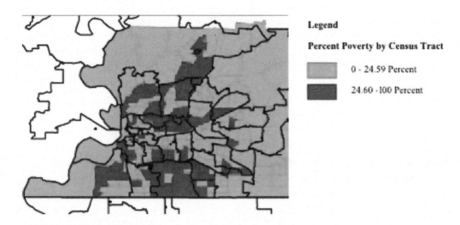

FIGURE 2.2
The average poverty rate at census tract level, 2015; red areas are at the average poverty rate or below, while green areas are above the mean poverty rate for the city.

offenses, arrests, and suspect residences and visualize and analyze DV hot spots, trends, and patterns in Memphis.

Review of the Literature

Several characteristics of DV are largely understood, while others remain in question. DV is a universal crime and remains fairly well distributed across neighborhoods and racial, ethnic, and social classes. However, certain factors play a role in (1) how law enforcement responds to DV, (2) whether victims are more likely to report experiences of DV victimization, and (3) how victims access services that may break the cycle of DV. GIS technology can assist in expanding this understanding to more comprehensively address this form of violence.

Law Enforcement Response to Domestic Violence

The law enforcement response to DV has shifted dramatically over the years. Historically, DV was handled informally, a strategy tied to the belief in the privacy of the family to handle matters within the home. In the wake of the Minneapolis Domestic Violence Experiment that prompted mandatory arrest policies for DV offenders (Sherman, Schmidt, and Rogan 1992), the response to DV has been more centralized within the criminal justice system. This response begins with law enforcement, often termed the "gatekeepers" of the criminal justice system.

Unfortunately, criminal justice handling of DV cases is often fraught with complications. Victims often report feeling that they are not taken seriously by police and prosecutors, especially if they have multiple encounters with law enforcement. They describe police officers at best as uncaring and dismissive and at worst as outright rude and abrasive (Carbone-Lopez, Slocum, and Kruttschnitt 2015; Stephens and Sinden 2000). DV victims may also face difficulty throughout court proceedings if they decide not to testify, recant their statements, or request leniency for their offender. These decisions can generate frustration from practitioners, particularly if their program success is reliant on the number of convictions that could result from victim participation in criminal justice processes (Raeder 2006; Riger, Bennett, Wasco, Schewe, Frohmann, Camacho, and Campbell 2002:96). Disparities in treatment for DV cases are also evident in the decision to prosecute; DV victims may face lower chances of prosecution for their offenders in cases of DV related to sexual assault (Campbell, Wasco, Ahrens, Sefl, and Barnes 2001). It is perhaps no surprise, then, that DV remains one of the most underreported crimes to police (Felson and Paré 2005).

Reporting Experiences of Domestic Violence Victimization

Research into how and why victims choose to report DV victimization to the police suggests that several factors may determine whether a victim is inclined to formally report victimization. Incidents involving sustained and more severe injuries, especially those that involve firearms, are more likely to trigger a call to the police (Bachman and Coker 1995; Bonomi, Trabert, Anderson, Kernic, and Hot 2014). Also, women who are married, those who have children in the home, and those who are minorities tend to be more likely to seek out assistance from the criminal justice system (Akers and Kaukinen 2009; Hutchinson and Hirschel 1998; Hollenshead, Dai, Ragsdale, Massey, and Scott 2006; Kaukinen, Meyer, and Akers 2013; Lipsky, Caetano, and Roy-Byrne 2009). In contrast, more informal assistance seems a more practical solution with victims who do not want to face the well-known frustration, confusion, and hassle of dealing with the criminal justice system (Fugate, Landis, Riordan, Naureckas, and Engel 2005). Income also may play a role; women in wealthier areas are more likely to want to avoid pursuing charges, as they may be able to call on other resources that lower-income women do not have access to, including medical care and readily available housing (Cerulli, Kothari, Dichter, Marcus, Kim, Wiley, and Rhodes 2015; Miller 1989). Finally, internal barriers, in terms of the victims themselves, may also affect the likelihood of reporting. Many victims of DV blame themselves for the violence, particularly due to the stigma of the violence (e.g., Overstreet and Quinn 2014), and especially in cases where victims are involved in multiple abusive relationships (e.g., Cui, Gordon, Ueno, and Fincham 2013). These factors could potentially diminish help-seeking and consequently reporting.

Access to Victims Services

Victims' access to social services and support for DV victimization can also be difficult. Despite difficulties in experiences with law enforcement, victims may choose to seek out social services because, as victims, they can access assistance in obtaining an order of protection, pursuing custody of minor children, and receiving emergency or temporary housing, among other things. Yet issues exist here, as well. Even when victims do end up seeking help from social services, that particular agency may not be aware of all the services available to victims of DV, or even whether all victims are eligible for these resources, hence providing limited relief (Fugate et al. 2005).

Additional concerns lie in the provision of services. Research increasingly recognizes that the likelihood of seeking formal (i.e., through the criminal justice system) versus informal (i.e., social support) help varies across racial and ethnic backgrounds. Black women are more likely to seek out formal resources by reporting their victimization to authorities and seeking out hospital services (Lipsky et al. 2009), while Hispanic and white women are more likely to rely on more informal DV services and family support (Hollenshead et al. 2006; but see Ingram 2007). However, the lack of culturally sensitive and appropriate services often fails to recognize diversity in help-seeking and the capacity to seek help, as well as the provision of intervention programming (Sumter 2006).

GIS Technology and Domestic Violence

GIS technology may offer a unique advantage in understanding DV in terms of criminal justice response, victim reporting, and victim access to services. GIS provides a visual display of crime data using points, lines, and polygons. Within criminological research, crime event locations are most commonly represented using points through geocoding processes. GIS has an advantage over standard maps like those stored in Google Maps because GIS is able to analyze locations in relation to spatial correlates of crime (Ratcliffe 2010), as well as individual-level characteristics such as weapon usage, victim demographics, and other factors (see, e.g., Iratzoqui and McCutcheon 2018). Studies on social disorganization have long theorized that crime is higher in areas with lower-quality, poorly constructed housing that tends to be built in large quantities in little space (e.g., Sampson and Groves 1989). This work has prompted the use of geospatial analysis to better understand the role of geography and the opportunity for crime. By connecting criminological research to criminal justice policy, law enforcement may be able to identify "hot spots" of crime, areas with high criminal activity that necessitate larger police presence. Common programs utilizing these technologies include CrimeStat©, which allows for more nuanced geospatial modeling to identify factors for why crime incidents are clumped within certain locations (Levine 2006).

One question left unanswered, however, is whether location and DV may be tied in similar ways, particularly within areas of concentrated disadvantage that have more areas of disrepair, may view violence as normalized, and have substantial alcohol outlets and limited education for residents (as summarized in Beyer, Wallis, and Hamberger 2015). Employing GIS to examine DV across these types of communities may more effectively focus DV prevention and response in both research and policy, through collaborative partnerships between researchers and practitioners (Murray et al. 2016).

GIS can also offer insight into understanding victim reporting patterns. If DV is more likely to occur in high-density areas like public housing (Raphael 2001), victimization is more likely to be detected by neighbors living close by. However, with higher numbers of resident turnover, victimization is less likely to be reported by residents or their neighbors who may not maintain social ties. Thus, the crime that gets reported may not be reflective of all DV (Murray et al. 2016). A better understanding of who reports victimization, particularly within areas already known to experience high rates of violence, is essential to identify unknown victims who may not feel equipped to handle the formal process of identifying and prosecuting their attackers. This is particularly true for victims who may still be financially and/or romantically dependent upon their offenders.

Additionally, GIS can highlight patterns of victim utilization of social services. By identifying areas of need, GIS can help pinpoint if certain populations are underserved, as well as the efficacy of service provision (Hetling and Zhang 2010). GIS can also provide a tool for victim service providers to coordinate services between agencies by targeting their services to specific locations within a community, city, or county (Stoe, Watkins, Kerr, Rost, and Craig 2003). It makes sense that DV victims may be more inclined to report if they live relatively near a location that has available victim services. Victims may be less aware of services that are not readily available within their neighborhoods, or if they are unable to obtain transportation to seek them out (Logan, Stevenson, Evans, and Leukefeld 2004). Thus, the location of services relative to where victims report experiencing DV seems especially relevant for further research.

Methodology

The current study uses data from multiple databases maintained by a large law enforcement agency (LEA) within Memphis, Tennessee, a municipality in the Southeastern United States, and covers incidents between October 1,

2017, and September 30, 2018. Calls for service (CFS) data provided the address of origin for each citizen call for assistance for "domestic disturbance" (n = 54,041). Calls were limited to only cases involving intimate partners (IPDV). Additional emphasis was placed on the violent crime of aggravated assault/domestic violence (AA/DV). Offense data provided the address of origin for each IPDV offense report (n = 11,189) and for each IPDV offense involving AA/DV (n = 1,367). Arrest data provided the residence address of each IPDV suspect arrested for AA/DV (n = 898). The LEA redacted names and other personally identifying information prior to sharing the data with the researchers.

It is important to note that CFS does not necessarily result in the production of an offense report and that offense reports do not necessarily result in an arrest. Data on offenses are produced when an officer responds to a call and takes a report. Often, when officers are dispatched to origins of CFS, no one is there to provide a report. Arrest data are generated when a law enforcement officer makes a physical arrest or an arrest warrant is issued. An arrest may happen at any time and is not dependent upon a CFS or a current offense report. Consequently, the current data are a conservative estimate. That is, of the 898 IPDV suspects arrested for AA/DV during the period, it is unknown how many of them committed their offense during the period; suspects could have committed offenses in prior years and were arrested only much later.

Each set of addresses, including the address of the offense and the call for service, was coded and mapped separately. Addresses were aggregated based on a ward map provided by the LEA. This LEA has nine (9) precincts, each comprising six (6) smaller subdivisions (wards). These smaller subdivisions are called "wards" and represent similarly sized areas of patrol for each precinct. Geographically, these areas do not tend to be congruent with census boundaries and are primarily used to delegate areas and tasks for patrol.

A street GIS file (i.e., shapefile) of the county within which the LEA is located was accessed through the U.S. Census TIGER/Line files and served as an address locator for all addresses. This task was necessary to geocode addresses and to associate the addresses with streets. Prior to geocoding, the addresses were evaluated for coding errors, which could have resulted in inaccurate locations and maps. Because thousands of officers populate the cases into the databases and input addresses and do not use a standard convention for doing so, many addresses do not initially match and are not mapped or are incorrectly mapped. Data were cleaned by researchers in order to decrease error, using techniques such as find commands and search and replace and at times determining address points through the utilization of online map platforms, such as Google Maps.

Proper coding was established using a threshold of accuracy. This allowed for the best candidate (address) to be paired with the most congruent geospatial location. Using this logic, if an address is a perfect match using the address locator it received a score of 100. For this project, a minimum match score of 85 was selected, which is nearer the minimum required for establishing "good matches."

Using the "select attributes" function, incidents were selected by case specifics, such as crime type. Specifically, those cases that are categorized as domestic disturbance are selected. Interval differences were categorized by the counts of crime by ward. The distribution or range within categories provided the ability to display these differences. The Jenks natural breaks classification method was used to create natural break points in the distribution to classify the ward counts into categories. The final maps depict differences by classification based on each count.

Another classification methodology used the standard deviation of the distribution to classify crime counts. Categories under this classification scheme included interval ranges of the proportion of standard deviation from the mean. Lastly, a density/heat map was created for CFS for a domestic disturbance. Cases were limited to CFS because they included a sufficiently large number of cases for the analysis to be conducted. The mapping procedure used address point data to show the variation of kernel dot density, which can be observed by adjusting the color feature as the brighter areas show areas denser with dots or addresses. Kernel density is calculated through the addresses (points) provided within the data. The density score is determined by the number of points within each raster cell. Once calculated, the density calculation provides a continuous incident distribution. Kernel density allows for density smoothing by the search radius of the kernels. For the current analyses, the planar method used an output cell size of 0.0014, with the square map unit as the area unit.

Analytical Technique

Once maps were created and data aggregated, the data were then compared to determine variation by ward ($n = 54$). Maps and the new geo-databases were then compared for consistencies to determine the wards with the highest counts of DV, which included counts and rates of the various measures of DV described above. Finally, a list of wards and ranking was created based on the consistency of wards across databases being toward the top of DV counts.

Results

Figure 2.3 depicts a count map of CFS for domestic disturbance, using red and orange coloring to depict areas with the highest numbers of calls and blue and green to depict areas with the lowest numbers of calls. Yellow represents areas with counts between the highest and lowest areas. Of the 54

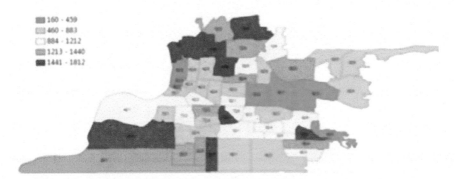

FIGURE 2.3
A count map of calls for service for domestic disturbance, from Memphis Police Department data, by census tract.

wards, 17 (31.5%) had more than 1,212 calls during the 12-month period, 9 wards (orange) had between 1,213 and 1,440, and 8 wards (red) had more than 1,440 calls (i.e., more than 120 calls per month). This map clearly indicates a concentration of calls in the northern and southwestern wards. In fact, five of the six wards in both the northern precinct (121, 122, 123, 124, and 125) and the southwestern precinct (221, 222, 223, 225, and 226) had among the highest counts of CFS. It is important to remember, however, that this map depicts simple counts. Larger, more populated areas will naturally have more CFS because they have more people. This is true for the northern and southwestern precincts; they are the more populous precincts.

Figure 2.4 also uses the CFS count, with a standard deviation map created relative to the mean CFS count. As shown, the red and orange areas have counts well above the average, while the blue and green areas have counts well below the average. Yellow areas are very close to the mean. Although four wards had numbers of CFS 1.5 or more standard deviations below the

FIGURE 2.4
Standard deviation of calls for service by census tract using Memphis Police Department data.

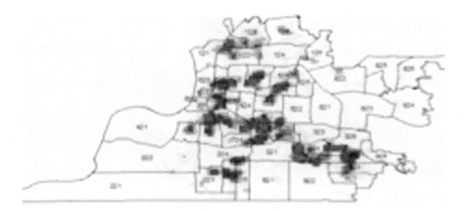

FIGURE 2.5
Kernel density ("heat") map of calls for service for domestic disturbance using Memphis Police
Department data.

mean, two of those wards are very small, with smaller populations relative
to other wards. Similar to Figure 2.3, the three wards with counts significantly higher than the mean (121, 226, 426) were again in the northern and
southwestern segments of the city.

Figure 2.5 is a kernel density ("heat") map of CFS for domestic disturbance. The map shows data clustered into a candy cane shape, running from
northwest to southeast with two additional clusters, one to the north and one
to the south. The straight part of this area approximates a city street known
to be a high-crime corridor.

Figures 2.6 and 2.7 specifically focus on offenses coded as IPDV by law
enforcement. Both the count map (Figure 2.6) and the standard deviation

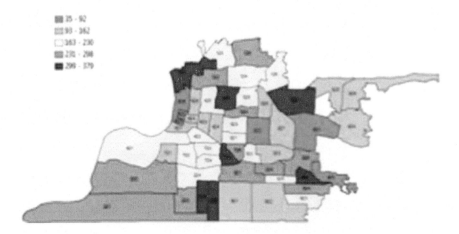

FIGURE 2.6
Count of offenses coded as intimate partner domestic violence by ward using Memphis Police
Department data.

FIGURE 2.7
Standard deviation of intimate partner domestic violence by ward using Memphis Police Department data.

map (Figure 2.7) are congruent with the CFS maps. One clear difference, however, is that the standard deviation map for IPDV offenses indicates that five wards were 1.5–2.1 standard deviations above the mean. Three of these five mirrored the CFS standard deviation map (121, 226, and 426). An additional two (725 and 822) were significantly higher than average in terms of IPDV offenses. The four wards whose standard deviations were significantly below the mean (522, 622, 623, 823) remained constant when compared to the CFS standard deviation map.

Figure 2.8 presents a kernel density map of IPDV offenses by ward. Similar to Figure 2.5, the shape resembles a hook or candy cane but identifies more strongly clustered offenses along the high-crime corridor. In this case, this pattern suggests the presence of multi-family housing units in areas where offense reports cluster, because families themselves are clustered.

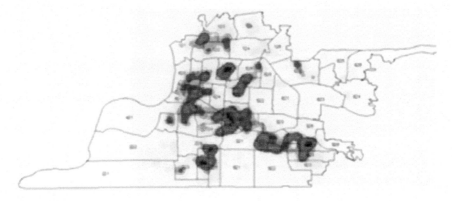

FIGURE 2.8
Kernel density map of intimate partner domestic violence offenses, with Memphis Police Department data, by ward.

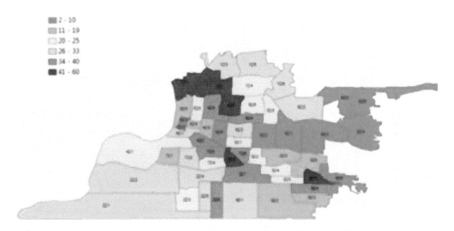

FIGURE 2.9
The numbers of intimate partner domestic violence filtered by aggravated offenses using Memphis Police Department data at ward level.

In addition to all IPDV offenses, maps were created using only IPDV offenses involving the violent crime of aggravated assault/domestic violence. Figure 2.9 indicates that the numbers of IPDV involving AA/DV were highest in the same (i.e., northern) wards that both IPDV and CFS were highest. However, this area also has the most people and a significant concentration of multi-family housing units. The dark orange and red wards along the high-crime corridor once again reflect problems in these wards; not only do these areas have a lot of CFS for domestic disturbance and IPDV offenses, but the offenses are more likely to involve a violent crime.

Figure 2.10 indicates that these results can be more refined. Although the standard deviation map of IPDV offenses involving AA/DV indicates that

FIGURE 2.10
A standard deviation ward map of mean intimate partner domestic violence offenses involving aggravated assault using Memphis Police Department data.

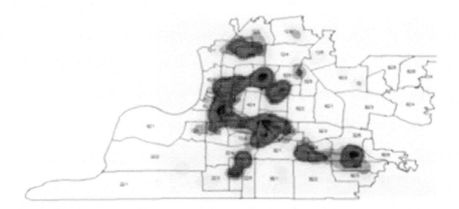

FIGURE 2.11

A kernel density map of intimate partner domestic violence offenses involving aggravated assault using Memphis Police Department data.

several wards have means whose standard deviations are significantly above average (as indicated by the orange and dark orange areas), only one ward has a mean more than 2.5 standard deviation units above average. It should be noted that the number per ward ranged from 2 to 60, which is due to the exclusivity of the definition of IPDV.

Figure 2.11 reports a kernel density map that more clearly illustrates the cluster pattern of IPDV offenses involving AA/DV. The dark purple areas indicate areas with the highest concentrations of IPDV involving AA/DV. Although several clusters are centralized squarely within one ward (i.e., 426, 925), many clusters span several wards. This has significant implications for developing place-based strategies led by law enforcement since some of these clusters even span multiple precincts. In addition to examining CFS, IPDV, and IPDV offenses involving AA/DV, the residential addresses of individuals arrested for AA/DV (not just IPDV) were mapped. Given the lower number of arrestees, however, the standard deviation maps were not constructed.

Figure 2.12 reports a count map based on offender residences. As in previous maps, areas with the fewest indicators of a DV problem remained constant (blue areas) and the residences of individuals arrested for AA/DV were, generally, in the problem areas. The two areas with the highest numbers of individuals arrested for AA/DV were wards 221 and 524, both with between 20 and 28 arrests for AA/DV during the period.

Figure 2.13 plots out the IPDV rate at census tract level per 100,000 population with poverty. The population and poverty data are gathered from the U.S. Census. As shown and discussed previously the map demonstrably shows the highest IPDV rates overlap areas that have higher poverty rates, with less of a pattern as rates decline.

When these maps are examined as a whole, clear patterns emerge. The candy cane distribution creates a semi-circle around areas of the city less plagued by DV (wards 423, 424, 522, 821, 823, 323, 326) and follows a distinct line from

FIGURE 2.12
A count map of the residence of individuals arrested for aggravated assault by ward.

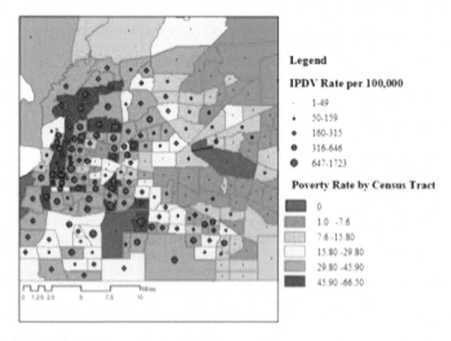

FIGURE 2.13
Multivariate map of poverty projected under the intimate partner domestic violence rate per 100,000 at census tract level.

northwest to southeast along a well-known high-crime corridor. While ward 121 has consistently high numbers, it also is one of the largest and most populous wards in the city. Ward 426 seems to be consistently high in all the maps and is a consistent cluster center. Wards 524 and 925 also seem to be hubs of high DV activity. Aside from these clear and centrally located cluster hubs, many clusters span multiple wards in a precinct and even multiple precincts, which have implications for law enforcement in developing responses if law enforcement must function according to the ward and precinct boundaries. On the other hand, it is likely that many of the clusters are connected to multi-family housing units which may facilitate the development and implementation of programs to address DV, generally, and IPDV specifically.

Conclusion and Recommendations

GIS technologies are further able to enhance relationships between researchers and practitioners. The current analyses identify certain clustering or "hot spots" of domestic violence within the metropolitan area, many of which replicate already commonly known areas of high crime. Like other violent crimes, IPDV tends to be clustered within several high-occupancy areas, with only a few high-crime areas reporting the most violent IPDV offenses (i.e., those involving AA/DV). As demonstrated by the above analysis, several clusters of the AA/DV offenses span several wards, though there is some overlap between areas that are higher in poverty and domestic violence. However, this pattern is not consistent, as some areas with lower poverty rates do exhibit higher domestic violence rates and vice versa. That said, these findings provide some guidance for law enforcement, particularly practitioners, moving forward.

It is important to note that any defined boundaries, including census tracts and wards, are grouping individuals who may not be consistent with the aggregate-level patterns, as it comes to socio-economic and socio-cultural factors. More economic and structural data are available at the census tract level as compared to the wards as defined by the local police department. This is important to consider as measures and relationships may vary based on the level of analysis. At the same time, once groups, cultures, are aggregated together, much is lost; this will continue to be a limitation in all macro-level research. It is crucial from the data we gather that we continue examining such data at multiple levels of analysis so that various groups are considered and treatment is best applied using a focused lens.

The analyses in this chapter indicate that patterns of IPDV are more similar to other forms of violence, echoing prior research (Iratzoqui and McCutcheon 2018). Law enforcement strategies that generally focus on violence prevention can have a similar impact on rates of IPDV, as well. Yet, given the more widespread nature of AA/DV, perhaps more aggressive strategies are warranted. One recommendation moving forward for law enforcement is a continued and more extensive partnership with academics, whose familiarity with the incident-specific causes of DV may serve to fine-tune prevention efforts and more effectively reduce this form of violence.

A more comprehensive awareness of DV by law enforcement may also address victim likelihood of reporting DV. Increased sensitivity to the risk factors that face victims of DV may allow law enforcement to understand the particular difficulties victims face in choosing whether or not to report their experiences of victimization. Similarly, this understanding may also enhance victims' access to services, particularly in terms of law enforcement communication of their availability. If law enforcement agencies develop a more comprehensive understanding of DV, they will be more likely to tap into current available resources to which they can subsequently refer victims more successfully. Future research in this area can adapt the GIS technology toward more comprehensively evaluating patterns of DV, IPDV, and AA/DV, particularly in cities with large disadvantaged populations who are experiencing the highest rates of crime and are at the greatest need for victim services to break that cycle.

References

Akers, C., and C. Kaukinen. 2009. The police reporting behavior of intimate partner violence victims. *Journal of Family Violence* 24:159–71.

Bachman, R., and A. L. Coker. 1995. Police involvement in domestic violence: The interactive effects of victim injury, offender's history of violence, and race. *Violence and Victims* 10:91–106.

Beyer, K., A. B. Wallis, and L. K. Hamberger. 2015. Neighborhood environment and intimate partner violence: A systematic review. *Trauma, Violence, and Abuse* 16:16–47.

Bonomi, A. E., B. Trabert, M. L. Anderson, M. A. Kernic, and V. L Holt. 2014. Intimate partner violence and neighborhood income: A longitudinal analysis. *Violence Against Women* 20:42–58.

Burraston, B., J. C. McCutcheon, and S. J. Watts. 2018. Relative and absolute deprivation's relationship with violent crime in the United States: Testing an interaction effect between income inequality and disadvantage. *Crime & Delinquency* 64:542–60.

Campbell, R., S. M. Wasco, C. E. Ahrens, T. Sefl, and H. Barnes. 2001. Preventing the "second rape": Rape survivors' experiences with community service providers. *Journal of Interpersonal Violence* 16:1239–59.

Carbone-Lopez, K., L. A. Slocum, and C. Kruttschnitt. 2015. "Police wouldn't give you no help": Female offenders on reporting sexual assault to police. *Violence Against Women* 22:366–96.

Cerulli, C., C. Kothari, M. Dichter, S. Marcus, T. K. Kim, J. Wiley, and K. V. Rhodes. 2015. Women experiencing intimate partner violence: Do they forgo the criminal justice system if their adjudication wishes are not met? *Violence and Victims* 30:16–31.

Cui, M., K. Ueno, M. Gorodon, and F. D. Fincham. 2013. The continuation of intimate partner violence from adolescence to young adulthood. *Journal of Marriage and Family* 75:300–13.

Felson, R. B., and P. P. Paré. 2005. The reporting of domestic violence and sexual assault by nonstrangers to the police. *Journal of Marriage and Family* 67:597–610.

Fugate, M., L. Landis, K. Riordan, S. Naureckas, and B. Engel. 2005. Barriers to domestic violence help seeking implications for intervention. *Violence Against Women* 11:290–310.

Hetling, A., and J. Zhang. 2010. Domestic violence, poverty, and social services: Does location matter? *Social Science Quarterly* 91:1144–63.

Hollenshead, J. H., Y. Dai, M. K. Ragsdale, E. Massey, and R. Scott. 2006. Relationship between two types of help seeking behavior in domestic violence victims. *Journal of Family Violence* 21:271–79.

Hutchison, I. W., and J. D. Hirschel. 1998. Abused women help-seeking strategies and police utilization. *Violence Against Women* 4:436–56.

Ingram, E. M. 2007. A comparison of help seeking between Latino and non-Latino victims of intimate partner violence. *Violence Against Women* 13:159–71.

Iratzoqui, A., and J. McCutcheon. 2018. The influence of domestic violence in homicide cases. *Homicide Studies* 22:145–60.

Kaukinen, C. E., S. Meyer, and C. Akers. 2013. Status compatibility and help-seeking behaviors among female intimate partner violence victims. *Journal of Interpersonal Violence* 28:577–601.

Levine, N. 2006. Crime mapping and the Crimestat program. *Geographical Analysis* 38:41–56.

Lipsky, S., R. Caetano, and P. Roy-Byrne. 2009. Racial and ethnic disparities in police-reported intimate partner violence and risk of hospitalization among women. *Women's Health Issues* 19:109–118.

Logan, T. K., E. Stevenson, L. Evans, and C. Leukfeld. 2004. Rural and urban women's perceptions of barriers to health, mental health, and criminal justice services: Implications for victim services. *Violence and Victims* 19:37–62.

Miller, S. L. 1989. Unintended side effects of pro-arrest policies and their race and class implications for battered women: A cautionary note. *Criminal Justice Policy Review* 3:299–317.

Murray, C., R. Bunch, and E. D. Hunt. 2016. Strengthening community-level understanding of and responses to intimate partner violence using geographic information systems (GIS). *Journal of Aggression, Conflict and Peace Research* 8:197–211.

Overstreet, N. M., and D. M. Quinn. 2014. The intimate partner violence stigmatization model and barriers to help seeking. *Basic and Applied Social Psychology* 35:109–22.

Peek-Asa, C., A. Wallis, K. Harland, K. Beyer, P. Dickey, and A. Saftlas. 2011. Rural disparity in domestic violence prevalence and access to resources. *Journal of Women's Health* 20:1743–49.

Raeder, M. S. 2006. Domestic violence in federal court: Abused women as victims, survivors, and offenders. *Federal Sentencing Reporter* 19:91–104.

Raphael, J. 2001. Public housing and domestic violence. *Violence Against Women* 7:699–706.

Ratcliffe, J. 2010. Crime mapping: Spatial and temporal challenges. In *Handbook of quantitative criminology*, ed. A. R. Piquero and D. Weisburd, 5–24. Springer, New York.

Riger, S., L. Bennett, S. M. Wasco, P. A. Schewe, L. Frohmann, J. M. Camacho, and R. Campbell. 2002. Evaluating services of domestic violence and sexual assault. *Sage series on violence against women*. SAGE, Thousand Oaks, CA.

Sampson, R. J., and W. B. Groves. 1989. Community structure and crime: Testing social-disorganization theory. *American Journal of Sociology* 94:774–802.

Sherman, L. W., J. D. Schmidt, and D. P. Rogan. 1992. *Policing domestic violence: Experiments and dilemmas.* Free Press, New York, NY.

Stephens, B. J., and P. G. Sinden. 2000. Victims' voices: Domestic assault victims' perceptions of police demeanor. *Journal of Interpersonal Violence* 15:534–47.

Stoe, D., C. R. Watkins, J. Kerr, L. Rost, and T. Craig. 2003. Using geographic information systems to map crime victim services. National Institute of Justice, U.S. Department of Justice, Washington, DC. http://www.ojp.usdoj.gov/ovc/public ations/infores/ geoinfosys2003/welcome.html.

Sumter, M. 2006. Domestic violence and diversity: A call for multicultural services. *Journal of Health and Human Services Administration* 29:173–90.

Truman, J. L., and R. E. Morgan. 2016. *Criminal victimization, 2015.* Bureau of Justice Statistics, U.S. Department of Justice, Washington, DC.

3

Geospatial Approaches to Intimate Partner Violence

Alison M. Pickover and J. Gayle Beck

CONTENTS

Introduction

Gender-based violence is a pervasive, global problem (World Health Organization [WHO] 2012, 1). For many years, gender-based violence was attributed to characterological factors (e.g., perpetrator's poor impulse control) or cultural norms (e.g., power inequities in patriarchal societies) (Heise 1998, 262–290). However, more recently, researchers have taken a broader perspective on gender-based violence, reconsidering both its precipitants and how prevention and intervention programs are designed. Modern frameworks emphasize the impact of individual-level factors (e.g., one's socio-demographic features and experiences), as well as the immediate and broader environment (e.g., one's relationships, community, and society). These frameworks suggest that individuals and environments influence one another, and that to fully understand and address gender-based violence, one must examine its relational, geospatial, and cultural contexts.

This chapter will examine geographic factors that are associated with the most-commonly experienced form of violence against women: intimate partner violence (IPV; WHO 2012, 1). We will look at the presence of this problem, including its scope and prevalence, and the reasons why these figures can be hard to determine. Then, we will explore the geographic correlates of IPV and introduce geographic information systems (GIS) technology as a tool for studying it. We will provide examples of studies that have integrated GIS technology and the study of IPV, including some of our own work, to demonstrate the ways in which this technology can be used and critical considerations when using it. Finally, we will speak to the future of this body of research, how it can grow, and why now is an important time for such work.

What Is Intimate Partner Violence? Definition and Scope

Intimate partner violence has been defined by the Centers for Disease Control as "physical violence, sexual violence, stalking, and psychological aggression (including coercive tactics) by a current or former intimate partner (e.g., spouse, boyfriend/girlfriend, dating partner, or ongoing sexual partner" (Breiding et al. 2015, 11). Within the United States, the National Intimate Partner and Sexual Violence Survey (NISVS) indicated that one in four women report the experience of severe physical abuse, 16% report sexual violence, and 47% of women experience psychological aggression, including humiliating and controlling behaviors, at the hands of an intimate partner during their lifetime (Smith et al. 2017, 1–2). Global data are a bit more difficult to collect; the WHO (2013, 2) reported that 30% of all women who have been involved in a romantic relationship have experienced physical and/or sexual violence from their intimate partner. Estimates of IPV vary geographically, with prevalence rates ranging from 23.2% in high-income countries to 37.7% in areas that the WHO defines as the Southeast Asia region (WHO 2013, 17). These data suggest that no matter which part of the world you examine, IPV will be present and will impact the women who live there.

IPV can be difficult to track, for several reasons. First, violence between romantic partners typically happens behind closed doors, unlike other forms of crime. Victims of IPV frequently do not reach out to law enforcement (Greenfeld et al. 1998, 17), fearing retribution from their partner or perceiving that the police cannot intervene in a meaningful way. Many victims of IPV also blame themselves for the violence (e.g., Reich et al. 2015, 1503), which can diminish help-seeking. Second, most police records track IPV by incident, not by person. Some women experience IPV in sequential romantic relationships (e.g., Cui et al. 2013, 300–313), yet current methods of tracking IPV do not allow consideration of the number of abusive partners that a woman may experience over time. Third, some women seek help only from grass-roots agencies for safety and shelter. These cases of IPV would typi-cally never appear in official reporting statistics, given the source of help-seeking and, possibly, the movement across state lines to ensure safety. For these reasons, indexes of IPV that rely on women seeking formal interven-tion obscure the scope of the problem. As an example, counts of domestic violence misdemeanor records and protective orders in the United States, compiled by the National Instant Criminal Background Check System, reg-ister as low as zero in some states. Mapping the frequency of these counts state to state (figures 3.1 and 3.2) illustrates their discordance sometimes with one another and with IPV metrics obtained using less-biased methodologies (e.g., the NIPSVS, an ongoing, random digit dial telephone survey; figure 3.3). In sum, because of tracking issues, available statistics on IPV should be con-sidered "best guesses."

Despite uncertainty about the prevalence of IPV, available evidence is very clear that IPV can have wide-ranging negative consequences for women. Current data suggest that over 40% of American women who experience IPV will also experience physical injury that is related to the relationship violence (Breiding et al. 2014, 41). Over 40% of women who are homicide victims in the United States are killed by their romantic partner (Cooper and Smith 2011,

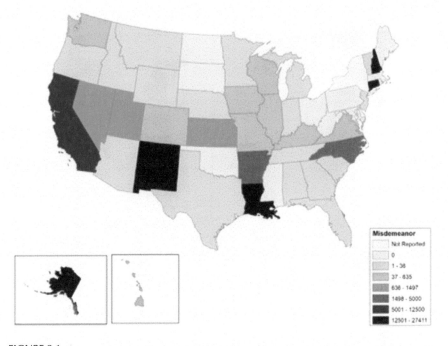

FIGURE 3.1
Domestic violence misdemeanor records submitted to the United States National Instant Criminal Background Check System Index. Data compiled by the National Coalition Against Domestic Violence (https://ncadv.org/state-by-state); data not available for New York state.

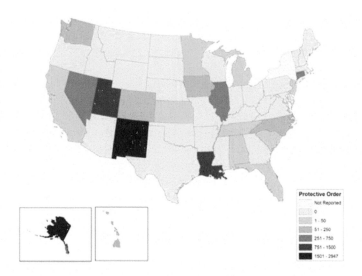

FIGURE 3.2
Protective orders submitted to the United States National Instant Criminal Background Check System Index. Data compiled by the National Coalition Against Domestic Violence (https://ncadv.org/state-by-state); data not available for New York state.

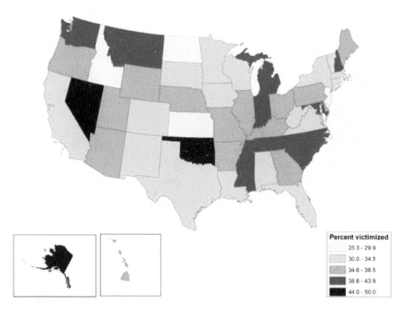

FIGURE 3.3
Lifetime prevalence on intimate partner violence among women in the United States. Reproduced from the 2010 National Intimate Partner and Sexual Violence Survey Summary Report, Table 7.4, Lifetime Prevalence of Rape, Physical Violence, and/or Stalking by an Intimate Partner by State of Residence – U.S. Women, NISVS 2010 (https://www.cdc.gov/violenceprevention/pdf/nisvs_report2010-a.pdf).

18). In general, survivors of trauma have been shown to have an increased risk for physical health problems, particularly cardiovascular conditions (e.g., Schnurr et al. 2007, 406–424). Mental health conditions can be anticipated following IPV, in particular, post-traumatic stress disorder (PTSD), depression, alcohol and drug abuse, and suicidal thinking (e.g., Golding 1999, 99–132). Given the extent and nature of negative consequences that can follow from exposure to IPV, it is imperative that we begin to understand this form of gender-based violence in context.

Understanding Intimate Partner Violence

Systems Perspectives

One of the most influential frameworks for conceptualizing IPV is the nested ecological theory of IPV (Ali and Naylor 2013, 616–617). This theory takes a systems approach to understanding the perpetration of violence against women. It emphasizes multiple factors and contexts, including personal history, relational context, social structures, societal institutions, norms, and cultural factors. For example, the model invokes person-level characteristics, such as having witnessed violence in the home during childhood, and situational or relational characteristics, such as male dominance and control of wealth in the family. The model further considers community characteristics such as neighborhood unemployment rate and isolation of the woman and family, and macroscopic factors such as cultural norms related to masculinity (Heise 1998, 262–290).

With this comprehensive approach to understanding IPV, a particular interest has arisen in the geographic context of IPV. Understanding the geographic distribution of IPV may aid clinicians, researchers, administrators, and policymakers in prevention efforts and guide decisions regarding advocacy resource allocation and geographic placement. It also may reveal areas in need of community outreach and intervention. Further, interventions might be enhanced by an ecologically informed approach to treatment (DePrince Buckingham and Belknap 2013, 259; Pickover et al. 2018, 13) – for instance, geographic factors might inform safety and escape-planning or treatment modality.

Geographic Factors

Two reviews (Beyer et al. 2015, 16–47; Pinchevsky and Wright 2012, 112–132) have usefully summarized the extant literature on geographic factors and IPV. Pinchevsky and Wright (2012, 112–132) specifically focused on known contributors to street violence and examined such variables as they contribute to violence in the home. They focused on six constructs: concentrated disadvantage (including aspects of neighborhood poverty such as crime, urbanicity, low income, and other economic deprivation variables; Wilson 1987, 283); collective efficacy (the extent to which neighborhoods are socially cohesive and neighborhood residents are willing to intervene on behalf of their community members) (Sampson et al. 1997, 918–919); ethnic heterogeneity; social ties; residential stability; and cultural norms. Broadly, the studies reviewed indicated associations between IPV and concentrated disadvantage, lower immigrant concentration, fewer social ties, and lower collective efficacy. Pinchevsky and Wright further showed the important buffering effects of social cohesion and support on IPV. Specifically, neighborhood density of friendships and common cultural backgrounds and norms may buffer against IPV victimization, although neighborhood familial ties may not be so protective (Wright and Benson 2010, 494). Beyer et al. (2015, 16–47) examined a broader range of studies. In their systematic review of 36 studies, 30 supported an association between neighborhood factors and physical or sexual IPV. Neighborhood-level socio-economic factors were most consistently linked to IPV, whereas mixed results were found for neighborhood disadvantage and community crime and violence. In both reviews, individual-level factors did not fully account for the relationships between geographic factors and IPV.

An important takeaway from the reviews of geographic factors and IPV is that dyadic interactions, which often only occur in the privacy of one's household, can be influenced by extradyadic neighborhood conditions, communal concerns, and community-driven efforts. This notion is echoed in the findings of research efforts that have focused on IPV-related outcomes. For example, Brown et al. (2005, 1478–1494) asked 90 African American women to complete questionnaires regarding their experiences of IPV and exposure to community violence, including directly experiencing, witnessing, or having knowledge of community violence, in the past three years. The researchers found an additive effect of IPV and community violence on self-reported post-traumatic stress symptoms, such that greater exposure to both forms of violence was associated with greater symptomatology.

Research has also incorporated assessment of perceptions of the geographic environment in examining IPV perpetration and mental health outcomes in its victims (Beeble et al. 2011, 287–306; Pickover et al. 2018, 1–22). This has primarily been achieved through the use of the perceived neighborhood disorders scale (PNDS; Ross and Mirowsky 1999, 412–432), a 13-item measure of perceptions of disorder in one's neighborhood. The scale assesses four dimensions of neighborhood order or disorder: physical order (e.g., neighborhood cleanliness); physical disorder (e.g., vandalism); social order (e.g., perceived safety); and social disorder (e.g., crime). In the first study to examine the effects of perceived neighborhood disorder on the well-being, Beeble et al. (2011, 287–306) examined depressive symptoms and quality of life over two years among 160 mothers with a history of physical victimization who survived IPV. They found that IPV frequency, perceived neighborhood disorder, and fear of additional victimization predicted baseline depressive symptoms, although perceived disorder was not associated with change in depression over time. Similar results were found for quality of life. In another study, Pickover et al. (2018, 1–22) examined depressive symptoms, hopelessness, and suicidal intent in a sample of 67 low-income African American women survivors of IPV. Perceived disorder and social support, differentiated by source (friends or family), interacted to predict mental health outcomes. Support from family members buffered against the negative effects of neighborhood disorder on depressive symptoms and hopelessness; support from friends buffered against the negative effects of neighborhood disorder on hopelessness and suicide intent. Thus, in addition to confirming the relation between perceived disorder and mental health following IPV, this study illuminated a potential conditional nature of the relationship.

Limitations of the Literature

Despite advances made in understanding the geographic factors that contribute to, or buffer against, IPV perpetration and mental health outcomes, researchers have noted limitations to the literature base. One particularly relevant limitation is the use of self-report measures of geographic factors. Mood and psychopathology (e.g., depression, anxiety; Kizilbash et al. 2002, 57–67) may influence memory and cognition and lead to erroneous or biased reports of neighborhood disorder, compromising the validity of self-report measures. This may be of particular concern, given high rates of depression, anxiety, and PTSD in survivors of IPV. Further, developmental factors may influence perception. In one study, adolescent dating violence was significantly associated with adolescent perceptions of neighborhood factors, but not with adult perceptions (Rothman et al. 2011, 207). This work speaks to the importance of stand-alone or adjunctive, objective measurements of geographic factors in IPV research.

A second concern regards the way in which the geographic environment is defined or quantified. In the past, when not self-defined by participants, researchers have often relied on census tracts to define neighborhoods. However, census tracts are often heterogeneous in size and create artificial boundaries. Analyses that rely on such boundaries may inaccurately represent an individual's most immediate surroundings (DePrince et al. 2013, 260–261). They may miss geographical features at the edge of a census tract that impact individuals in adjacent tracts. As a result, census tract boundaries may arbitrarily demarcate the environments of women whose realities are very similar. On the other hand, they may group together women who are spatially proximal to one another but exist in very different landscapes. A more refined approach to defining the neighborhood or environment could foster more meaningful and precise analyses.

Using GIS in IPV Research

GIS technology can accommodate significant flexibility in organizing geographic data and is a powerful tool for research that traverses the social and natural sciences. GIS allows for innovative approaches to measuring individuals' environments and offers the opportunity for improved analyses that address the shortcomings of previous studies. In this section, we will explore the application of GIS to three distinct yet related research topics: geospatial landscape of IPV; geospatial distribution of resources; and geographic factors associated with IPV mental health outcomes.

Geospatial Landscape of IPV

Most commonly, GIS technology has been used in IPV studies examining associations between geographic features of the environment (e.g., unemployment rate; residential stability) and georeferenced incidents of IPV. Facilitating such research is the relative ease with which geographic data can be obtained. For instance, U.S. census data is publicly available and downloadable in a format that easily lends itself to analysis. Often, researchers have several options regarding the level of aggregation of such data, allowing for fine-grained analysis of the environment. U.S. census data are available in various increments other than census tract; block-level and block-group-level data are the smallest units for which census data can be obtained by the general public. Researchers may also collect their own geographic data or collaborate with community or other macro-level entities (e.g., police departments to obtain crime data).

An example of GIS-facilitated geospatial analysis comes from Li et al. (2010, 531–539), who used multilevel modeling to investigate individual, household, and neighborhood factors associated with IPV among a particularly high-risk group, pregnant women, in Jefferson County, Alabama. They linked the following geocoded data at the census-tract level: 2,887 participant addresses, percentage unemployment, percentage African American, percentage households below poverty line, on public assistance with a single parent (concentrated disadvantage), percentage of households staying in the same residence at least five years (residential stability), and annual violent crime. Among other associations, Li et al. found links between physical and sexual violence and residential stability, even after accounting for individual and household factors.

Similar approaches have been used in even finer-grained analysis. Using Bayesian spatial modeling to accommodate small-area data, Gracia et al. (2014, 866–882) examined associations between small-area variations in IPV and neighborhood factors such as socio-economic status (property values, residential instability) and percentage of immigrants (census data), ratings of public disorder and crime (derived from policing activity), and social and physical disorder (coded by trained raters) across census block groups in a single police district in Valencia, Spain. IPV incidence was quantified as count of granted protective orders (georeferenced to the specific addresses of the incidents motivating the orders) in census block groups. The researchers found associations between IPV, higher immigrant concentration, higher levels of public disorder and crime, and higher levels of physical disorder.

Cunradi et al. (2011, 191–200) took a different approach in their application of GIS to IPV research, analyzing specific features of the built environment. In addition to examining census-obtained variables, these authors used georeferenced data to investigate relations between IPV and alcohol retail establishments in Sacramento, California. GIS software was used to join addresses associated with police phone calls regarding couples' physical or verbal altercations, IPV crime records, alcohol outlet densities, and census block-group data. Greater alcohol outlet density was linked to both increased IPV calls and increased crime reports.

It is of note that all three studies employed statistical methods that accounted for *spatial autocorrelation*, a measure of similarity between nearby observations, accommodating the possibility that (IPV-exposed) individuals in one neighborhood may be more similar to one another than individuals in another neighborhood. Inattention to spatial autocorrelation has been raised as a particular problem plaguing this research domain (Pinchevsky and Wright 2012, 128). Yet while spatial autocorrelation is usually treated as a statistical issue, it may also offer insights into the geographic distribution of IPV. Does IPV cluster in specific locations? If so, this phenomenon is relevant to the placement of resources. At least one GIS-informed study found that among women who experienced and reported IPV to the police, those women who lived closer together, relative to those who lived farther from each other, experienced more similar levels of psychological aggression (DePrince et al. 2013, 264). The authors suggested that these results might be due to differences in bystander behavior – that is, bystanders in different neighborhoods may be more or less likely to condone such behaviors, which may in turn influence displays of aggression. However, bystander behavior was not measured in the study.

It is important to emphasize that the aforementioned studies all used IPV data reported to police. Unfortunately, only half of the episodes of violence perpetrated against women are reported to the police (Greenfeld et al. 1998, 17). Sample generalizability may be low if women in such neighborhoods (samples) were more likely to report IPV to police (perhaps because they thought successful intervention by authorities was likely). Relatedly, in some geographic areas, police response to domestic assaults may be prioritized or arrests may be more common (Helting Zhang 2010, 1154–1155), which could confound explanations of apparent IPV clustering. When examining spatial autocorrelation of IPV, methods of IPV incidence collection should be carefully considered. In some cases, reliance on police data alone may be problematic.

Geospatial Distribution of Resources

Another interesting domain of GIS-informed IPV research concerns the geospatial distribution of IPV resources. Studies in this arena may examine broad issues such as resources for IPV survivors in rural versus urban environments, or they might look at features of the built environment, like the location of police departments, social service agencies, victim advocacy centers, and facilities housing IPV intervention programs or mental health practitioners. Regarding the former, recent reviews (Beyer et al. 2015, 41) have noted a particular need for advancing frameworks for understanding geographic factors as they influence IPV in rural areas.

Applying GIS technology to research on geographic disparities in IPV and IPV resources, Peek-Asa et al. (2011, 1743–1749) addressed three questions: (1) does IPV prevalence, frequency, and severity differ by rurality of one's environment? (2) do IPV services vary across more rural versus more urban environments?; and (3) does distance to IPV resources vary for women living in these different environments? To answer these questions, they linked cross-sectional self-report IPV data from 1,478 women seeking elective abortions in Iowa with participants' zip codes and georeferenced addresses for domestic violence intervention programs. Rurality was determined based on widely used coding systems (e.g., Rural Urban Commuting Area, US Department of Agriculture Urban Influence Codes). In their sample, prevalence, frequency, and severity of physical and sexual IPV were higher among women in small or isolated rural areas as compared with urban or large rural areas. Significant disparities were observed in available services, for example, intervention programs per county were higher in urban counties. Urban counties had, on average, 13.8 shelter beds per county, whereas beds per county ranged from 3.3 to 4.7 in remote rural, rural, and suburban counties. Conversely, a large percentage of rural and remote rural county programs provided transportation, compared with only half of urban counties. Finally, disparities were seen in distance to programs. For example, over a quarter of women in small rural or isolated areas, but no women in urban and large rural areas, had to travel more than 40 miles to their nearest services.

In the case of the built environment, GIS software packages are particularly useful in allowing researchers to derive multiple geospatial indexes. For example, GIS technology can be used to determine the number of agencies located within a specific geographic location, whether agencies border identified spaces, and spatial distance between an agency and another specific location (e.g., a town center or specific individual). A study by Hetling and Zhang (2010, 1144–1163) examined the associations between georeferenced domestic violence agencies, town socio-economic characteristics, and IPV in 169 Connecticut towns and cities. IPV was operationalized as arrests related to domestic violence incidents and domestic violence assaults. Hetling and Zhang largely did not find an association between police-reported IPV and presence (or lack) of domestic violence agencies. Perhaps, other geospatial indicators of resource availability, such as travel costs (Higgs 2004, 123), might be more salient measures of disparities, particularly given the economic consequences of IPV for victims. Travel time and hours of availability are other useful indicators of the built environment that can be derived using GIS technology (Hyndman et al. 2001, 1599–1609). These variables may hold particular relevance for low-income women exposed to IPV, given less flexibility in scheduling medical appointments due to reliance on public transportation (which may be unreliable or increase the length of the trip substantially) or challenging work schedules (e.g., night shifts, multiple part time jobs). Finally, ambulance response time can be mapped using GIS technology (Peters and Hall 1999, 1551–1566), which may provide an indicator of responsivity to domestic assaults.

Geographic Factors Associated with IPV Mental Health Outcomes

GIS has rarely been applied to the study of geographic factors and IPV mental health outcomes. The first study to integrate the methodology was authored by DePrince et al. (2013, 258–271) and estimated the effects of geographic variables, in addition to individual- (e.g., age, race) and event-level factors (e.g., type and severity of IPV) on PTSD and depressive symptoms in 192 female survivors of IPV survivors who reported violence by a male perpetrator to the police. Suggesting that the conditions of survivors' most immediate surroundings might have the greatest impact on their mental health, the authors cast a computer-generated grid over the spatial area where participants lived, allowing them to assign participants' addresses and associated geographic characteristics to 1,000-square-foot grid pixels, which they termed *proximal environments*. PTSD was associated with higher levels of fear and a greater percentage of single father households in the proximal environment. Depression was found to be associated with higher levels of fear and buffered by the percentage of Latinos living in the proximal environment.

In our own work, we are using GIS technology to analyze mental health outcomes associated with neighborhood factors operationalized two ways: crime aggregated to the proximal environment (1,000-square-foot grid pixel) and crime aggregated at the precinct level. Our sample includes 229 women residents of Shelby County, Tennessee, who were assessed for PTSD, depression, and general anxiety disorder within a mental health research clinic for women IPV survivors. Participant addresses and georeferenced crime data were linked using ArcGIS software (ESRI, Inc. 2013). We derived proximal environment and precinct crime values using the ArcGIS fishnet tool and publicly available maps of Shelby County, Tennessee. In examining crime as we intend to, we hope not only to determine whether crime and IPV mental health outcomes are linked but to determine which geographic spaces (proximal or precinct) may be most impactful on IPV survivor's mental health. Including two units of analysis in the same investigation also can shed light on whether null results found in other geospatial IPV studies might be explained by a lack of precision in operationalizing the neighborhood or, conversely, too-narrowly defined neighborhoods.

Our initial analyses suggest that the environment most impactful on mental health outcomes may differ for White and Black women (Pickover 2017, 36–47). For example, it appears that precinct-level crime may have more impact on the well-being of Black IPV survivors, whereas crime in the proximal environment may have more impact on White women. Our early results also suggest that the effects of social support may only be apparent at certain levels of measurement. For instance, social support appears to moderate the effects of geographic factors measured at the precinct level but not the proximal environment. Finally, geographic factors may have greater relevance for depressive disorders than fear- and anxiety-based ones, although more research is needed to clarify specific effects and mechanisms.

Future Directions

GIS is a powerful tool for studying IPV and its consequences. Research at the intersection of these two fields is in its infancy, and there is considerable promise for advancing this work the near future. For instance, few studies have applied GIS to the study of psychological IPV, a prevalent form of violence against women that can have just as detrimental effects on well-being as physical or sexual violence. GIS might also be applied to the examination of IPV assessment in primary care settings, service efficacy, and cost-effectiveness of interventions. Determining the most relevant spaces to consider when assessing and addressing IPV, based on the characteristics of the victims and the perpetrators as well as their communities, also warrants attention.

Special considerations in future GIS analysis should be given to under-reporting of IPV and other crime to the police and mobility of IPV victims. Women may move or go into hiding and be unable or unwilling to report abuse and provide addresses. Both issues affect accurate georeferencing of IPV, and thus new methodologies are needed to improve ecological validity of IPV-GIS studies. Murray et al. (2016, 197–211) have laid out a series of steps to guide research on IPV and geographic factors. In addition to outlining conceptual issues (e.g., defining the neighborhood in an ecologically valid way), they discuss technical aspects of GIS analysis, including vector and raster-based analyses. Their guidelines may be particularly informative for those making their initial foray into this research domain.

Ultimately, advancing the IPV-GIS literature will require interdisciplinary and innovative partnerships between the natural sciences and the fields of public health, epidemiology, psychology, social work, and medicine. As new data are uncovered, this work will be of great value to policy decisions, such as government allocation of funds and distribution of resources within hospitals, communities, and law enforcement agencies. Collaborative, data-driven work with urban planners and administrators can help researches and policymakers bring their visions to fruition. Together, their efforts can result in improved safety and quality of care for vulnerable women.

References

Ali, Parveen Azam, and Naylor, Paul B. 2013. "Intimate partner violence: A narrative review of the feminist, social and ecological explanations for its causation". *Aggression and Violent Behavior*, 18(6): 611–619. doi:10.1016/j.avb.2013.07.009

Beeble, Marisa L., Sullivan, Cris M., and Bybee, Deborah. 2011. "The impact of neighborhood factors on the wellbeing of survivors of intimate partner violence over time". *American Journal of Community Psychology*, 47(3–4): 287–306. doi:10.1007/s10464-010-9398-6

Beyer, Kirsten, Wallis, Anne Barber, and Hamberger, L. Kevin. 2015. "Neighborhood environment and intimate partner violence". *Trauma, Violence, & Abuse*, 16(1): 16–47. doi:10.1177/1524838013515758

Breiding, Matthew J., Basile, Kathleen C., Smith, Sharon G., Black, Michele C., and Mahendra, Reshma. 2015. *Intimate partner violence surveillance: Uniform definitions and recommended data elements, Version 2.0*. Atlanta, GA: National Center for Injury Prevention and Control, Centers for Disease Control and Prevention. https://www.cdc.gov/violenceprevention/pdf/ipv/intimatepartnerviolence.pdf

Breiding, Matthew J., Chen, Jieru, and Black, Michele C. 2014. Intimate partner violence in the United States –2010. Atlanta, GA: National Center for Injury Prevention and Control, Centers for Disease Control and Prevention. https://www.cdc.gov/violenceprevention/pdf/cdc_nisvs_ipv_report_2013_v17_single_a.pdf

Brown, Jorielle R., Hill, Hope M., and Lambert, Sharon F. 2005. "Traumatic stress symptoms in women exposed to community and partner violence". *Journal of Interpersonal Violence*, 20(11): 1478–1494. doi:10.1177/0886260505278604

Cooper, Alexis, and Smith, Erica L. 2011. *Homicide trends in the United States, 1980 – 2008*. Washington, DC: Bureau of Justice Statistics, NCJ 236018. https://www.bjs.gov/content/pub/pdf/htus8008.pdf

Cui, Ming, Ueno, Koji, Gordon, Melissa, and Fincham, Frank D. 2013. "The continuation of intimate partner violence from adolescence to young adulthood". *Journal of Marriage and Family*, 75(2): 300–313. doi.org/10.1111/jomf.12016

Cunradi, Carol B., Mair, Christina, Ponicki, William, and Remer, Lillian. 2011. "Alcohol outlets, neighborhood characteristics, and intimate partner violence: Ecological analysis of a California city". *Journal of Urban Health*, 88(2): 191–200. doi:10.1007/s11524-011-9549-6

DePrince, Anne P., Buckingham, Susan E., and Belknap, Joanne. 2013. "The geography of intimate partner abuse experiences and clinical responses". *Clinical Psychological Science*, 2(3): 258–271. doi:10.1177/2167702613507556

Golding, Jacqueline M. 1999. "Intimate partner violence as a risk factor for mental disorders: A meta-analysis". *Journal of Family Violence*, 14(2): 99–132. doi. org/10.1023/A:1022079418229

Gracia, Enrique, López-Quílez, Antonio, Marco, Miriam, Lladosa, Silvia, and Lila, Marisol. 2014. "Exploring neighborhood influences on small-area variations in intimate partner violence risk: A Bayesian random-effects modeling approach." *International Journal of Environmental Research and Public Health*, 11: 866–882. doi:10.3390/ijerph110100866

Greenfeld, Lawrence A., Rand, Michael R., Craven, Diane, Klaus, Partsy A., Perkins, Craig A., Ringel, Cheryl, Warchol, Greg, Maston, Cathy, and Fox, James Alan. 1998. "Violence by intimates: Analysis of data on crimes by current or former spouses, boyfriends, and girlfriends." https://bjs.gov/content/pub/pdf/vi.pdf

Heise, Lori L. 1998. "Violence against women an integrated, ecological framework". *Violence* Against Women, 4(3): 262–290. doi:10.1177/1077801298004003002

Hetling, Andrea, and Zhang, Haiyan. 2010. "Domestic violence, poverty, and social services: Does location matter?". *Social Science Quarterly*, 91(5): 1144–1163. doi:10.1111/j.1540-6237.2010.00725.x

Higgs, Gary. 2004. "A literature review of the use of GIS-based measures of access to health care services". *Health Services and Outcomes Research Methodology*, 5: 119–139. doi:10.1007/s10742-005-4304-7

Hyndman, Jilda C. G., Holman, C., and D'Arcy, J. 2001. "Accessibility and spatial distribution of general practice services in an Australian city by levels of social disadvantage". *Social Science & Medicine*, 53(12): 1599–1609. doi:10.1016/s0277-9536(00)00444-5

Kizilbash, Ali H., Vanderploeg, Rodney D., and Curtiss, Glenn. 2002. "The effects of depression and anxiety on memory performance". *Archives of Clinical Neuropsychology*, 17: 57–67. doi:10.1016/s0887-6177(00)00101-3

Li, Qing, Kirby, Russell S., Sigler, Robert T., Hwang, Sean-Shong, LaGory, Mark E., and Goldenberg, R. L. 2010. "A multilevel analysis of individual, household, and neighborhood correlates of intimate partner violence among low-income pregnant women in Jefferson County, Alabama". *American Journal of Public Health*, 100(3): 531–539. doi:10.2105/ajph.2008.151159

Murray, Christine, Bunch, Rick, and Hunt, Eleazer D. 2016. "Strengthening community-level understanding of responses to intimate partner violence using geographic information systems (GIS)". *Journal of Aggression, Conflict and Peace Research*, 8(3): 197–211. doi:10.1108/jacpr-01-2016-0209

Peek-Asa, Corinne, Wallis, Anne, Harland, Karisa, Beyer, Kristen, Dickey, Penny, and Saftlas, A. 2011. "Rural disparity in domestic violence prevalence and access to resources". *Journal of Women's Health*, 20(11): 1743–1749. doi:10.1089/jwh.2011.2891

Peters, Jeremy, and Hall, G. Brent. 1999. "Assessment of ambulance response performance using a geographic information system". *Social Science & Medicine*, 49(11): 1551–1566. doi:10.1016/s0277-9536(99)00248-8

Pickover, Alison Marisa. 2017. *An Ecological Systems Approach to Understanding Intimate Partner Violence Outcomes*. PhD diss., University of Memphis.

Pickover, Alison Marisa, Bhimji, Jabeene, Sun, Shufang, Evans, Anna, Allbaugh, Lucy J., Dunn, Sarah E., and Kaslow, Nadine J. 2018. "Neighborhood disorder, social support, and outcomes among violence-exposed African American women." *Journal of Interpersonal Violence*. Advance online publication. doi:10.1177/0886260518779599

Pinchevsky, Gillian M., and Wright, Emily M. 2012. "The impact of neighborhoods on intimate partner violence and victimization". *Trauma, Violence, & Abuse*, 13(2): 112–132. doi:10.1177/1524838012445641

Reich, Catherine M., Jones, Judiann M., Woodward, Matthew J., Blackwell, Nathali, Lindsey, Leslie, and Beck, J. Gayle. 2015. "Does self-blame moderate psychological adjustment following intimate partner violence?". *Journal of Interpersonal Violence*, 3(9): 1493–1510. doi: 10.1177/0886260514540800

Ross, Catherine E., and Mirowsky, John. 1999. "Disorder and decay: The concept and measurement of perceived neighborhood disorder". *Urban Affairs Review*, 34(3): 412–432. doi:10.1177/107808749903400304

Rothman, Emily F., Johnson, Renee M., Young, Robin, Weinberg, Janice, Azrael, Deborah, and Molnar, Beth E. 2011. "Neighborhood-level factors associated with physical dating violence perpetration: Results of a representative survey conducted in Boston, MA". *Journal of Urban Health*, 88(2): 201–213. doi:10.1007/s11524-011-9543-z

Sampson, Robert J., Raudenbush, Stephen W., and Earls, Felton. 1997. "Neighborhoods and violent crime: A multilevel study of collective efficacy". *Science*, 277: 918–924. doi:10.1126/science.277.5328.918

Schnurr, Paula P., Green, Bonnie L., and Kaltman, Stacey. 2007. "Trauma exposure and physical health." In *Handbook of PTSD: Science and Practice*, edited by Matthew J. Friedman, Terence M. Keane, and Patricia A. Resick, 406–424. New York: Guilford Press.

Smith, Sharon G., Chen, Jieru, Basile, Kathleen C., Gilbert, Leah K., Merrick, Melissa T., Patel, Nimesh, Walling, Margie, and Jain, Anurag. 2017. The National Intimate Partner and Sexual Violence Survey (NISVS): 2010 – 2012 State Report. Atlanta, GA: National Center for Injury Prevention and Control, Centers for Disease Control and Prevention. https://www.cdc.gov/violenceprevention/pdf/NISVS-StateReportBook.pdf

Wilson, William Julius. 1987. *The Truly Disadvantaged: The Inner city, The Underclass, and Public Policy*. Chicago, IL: University of Chicago Press.

World Health Organization. 2012. *Understanding and addressing violence against women*. Geneva: World Health Organization. https://apps.who.int/iris/bitstream/handle/10665/77432/WHO_RHR_12.36_eng.pdf?sequence=1

World Health Organization. 2013. Global and regional estimates of violence against women: Prevalence and health effects of intimate partner violence and non-partner sexual violence. Geneva: World Health Organization. https://apps.who.int/iris/bitstream/handle/10665/85239/9789241564625_eng.pdf?sequence=1

Wright, Emily M., and Benson, Michael L. 2010. "Immigration and intimate partner violence: Exploring the immigrant paradox". *Social Problems*, 57(3): 480–503. doi:10.1525/sp.2010.57.3.480

4

Gender Disparity and Economy in U.S. Counties: Change and Continuity, 2000–2017

Madhuri Sharma

CONTENTS

Introduction

There is an abundance of scholarship on spatial disparities of race-based under/over-representation across industries in the United States. Asians, for example, are over-represented in the technical fields, especially in the Silicon Valley area, whereas Hispanics in the southern and central parts of California over-represent in agro-industries and Blacks are clustered in the farms/plantations of the U.S. Southeast (Coe et al. 2012). A recent work suggests over-representation of Asians in technical and creative-class industries in the counties of Tennessee (Sharma 2018), whereas Blacks and Hispanics still dominate in agriculture, construction, and other ancillary sectors of economy (Sharma 2016). While Asians in Alabama are engaged in ship building and manufacturing (Lester and Nguyen 2015), in the case of North Carolina,

Latinos over-represent in their booming construction industry, with the state accounting a growth of 386% in Hispanics during 1990–2000 (Furuseth and Smith 2006). Other smaller towns and urban areas also attracted new minorities, most of whom provided labor in the fastest growing sectors of U.S. economy (Mohl 2007; Lichter and Johnson 2009). While a sizable scholarship has attributed to intellectual stimulation on changing social geographies, not much attention has been paid to the spatial changes pertaining to gendered differences in economy, especially due to a lack of high-quality data and specificity in analyzing and understanding gender economy. As of now, data on gender has to be gleaned from a variety of sources, which makes it tedious and unappealing to many social scientists.

Despite these difficulties, recently the American Fact Finder (AFF) and the American Community Survey (ACS) of the Bureau of U.S. Census have started providing data by gender for industries and occupations, along with their broad socio-economic and demographic characteristics that are crucial to understanding spatial patterns of gender representation across the counties of United States. This chapter specifically uses the five-year data estimates (2013–2017) of the ACS to answer simple and yet very important and timely research questions: What types of gender-based over/under-representations exist in various types of industries in U.S. counties? Which industries have shown significant improvement for women's participation during 2000–2017? How can one explain these spatial patterns using theories such as the human capital theory, neoclassical theory, and the labor segmentation theory? These are critical to our understanding of gender economy in current geopolitical environments. These analyses will provide insights into gender-based spatial differentiations that could help policymakers devise appropriate actions to reduce gender gaps in industries.

Background Contexts

Empirical Perspectives on Gendered Labor

Gender differences in economic outcomes have received growing attention from economists in recent times, and comprise an important and a positive step forward for gender equity (Schneebaum 2016; Whitmore and Nunn 2017). This is important since the globally factual gendered differences in incomes have been often attributed to work/industry-type differences and their performance spaces, which explain wide pay gaps across genders (Coe et al. 2012; HTTP7). On average, men are paid more than women for the same work (Coe et al. 2012; HTTP7; Robinson 2018; Sawhill 2016; Schneebaum 2016; Schuele and Louise 2018), and these are attributed to the idea that the tasks performed by women are conducted in private spaces – mostly dating back to women's primary roles in reproductive activities. In general, women still remain the most under-represented at every level in corporate America despite an increase in their sharing the work spaces with men. Despite their overall gains in educational attainments – with far greater shares of women earning college degrees compared to men now for almost 30 years – not much success has been noted for women getting into all varieties of industries (McKinsey & Company 2015; Krivkovich et al. 2017).

Women suffer disproportionately in the industries where both formal and informal work are performed largely in public spaces, whereas a major part of women's activities are performed in domestic, i.e., private spaces, inside the boundaries of their household, and hence becomes invisible and under-valued (Coe et al. 2012; Landau and Lewis 2018; McKinsey & Company 2015, Krivkovich et al. 2017; Reeves and Venator 2014), and this clearly is the dark side of reproductive and caring activities conducted largely by women. In reality, though, a majority of these activities, if performed by professionals, could comprise a major share of a household's income and a country's GDP (see Coe et al. 2012 on UK's analyses of unpaid work's monetary valuation). This occurs as they run through the opportunity costs of high child care costs, low wages, and the structurally insensitive tax programs, especially when one examines the economic challenges of raising a child in a single-parent household (Reeves and Venator 2014). For these very depressing reasons, the United States' current family leave policy is surprisingly equitable across genders – i.e., neither men nor women are guaranteed any paid leave, even though at an individual level, few states and organizations allow few weeks of family leave toward maternity benefits (Landau and Lewis 2018; Reeves and Venator 2014). These policies are equally unfair to all working parents, but these eventually pressurize the females into childcare such that the head of the household must not compromise with his "outside/public-space job/career" (Coe et al. 2012; Landau and Lewis 2018; Reeves and Venator 2014). These policies inadvertently restrict the educated, qualified, and skilled women, and certainly the single moms at far greater risks, from fully participating in the labor market. Such policies, along with historical and contextual divisions of gender labor, have in fact worked against women's fuller participation in industries that pay higher and have traditionally been male dominated (HTTP4).

A recent study showed substantial improvements for female participation in traditional male-dominated spheres, particularly manufacturing industries (HTTP3). This analysis suggests that some of the reasons that women consider while wanting to work in manufacturing include opportunities for challenging assignments, work–life balance, and attractive incomes – all of which comprise the most important aspects of a women's career. Many schools and manufacturing industries have improved training programs, and women are now trying to engage in typically male-dominated sectors such as construction/manufacturing and the like (HTTP2; HTTP3; HTTP4). In their survey of over 600 women in manufacturing and nearly 20 manufacturing executives to explore the effectiveness of manufacturing companies in attracting, recruiting, and retaining women, the Manufacturing Institute and Deloitte interviewed a number of manufacturing industry leaders, including male and female executives, prior STEP (Science Technology, Engineering, and Production) Ahead honorees, and STEP Ahead emerging leaders to inform overall insights. They found great talent among women to work in manufacturing and that women comprise one of U.S. manufacturing's largest pools of untapped talents. Women actually totaled 47% of the U.S. labor force in 2016, but only 29% of the manufacturing workforce comprised women (HTTP5; HTTP6; HTTP7), even though women earned more than half of all associate's, bachelor's, and master's degrees (HTTP5; HTTP6; HTTP8).

At an international scale, the United Nations (UN) has also conducted its own assessment of gender differences in industry employment which are critical to their fuller growth and participation (HTTP10; Whitmore and Nunn 2017). The most important elements of UN Women's economic empowerment include women's ability to participate equally in existing markets across industry types, their access to and control over productive resources, including access to decent work, control over their own time, lives and bodies, and increased voices in decision-making and in other production and work-related issues, that are currently missing because of their very small share of participation in male-dominated industries (HTTP10). These will eventually help them with meaningful participation in economic decision-making at all levels, from the household to international institutions. It is also important to recognize that education, skill development, and re-skilling over the life course to keep pace with rapidly changing technological and digital fields affecting the jobs are critical for women's overall economic health and social well-being as these get affected directly and indirectly from their income-generation opportunities and participation in the formal labor market (HTTP10; Whitmore and Nunn 2017).

Across the globe, as of now, almost 2.7 billion women are legally restricted from having the same choice of jobs as men due to cultural, political, and social reasons, and of the 189 economies assessed in 2018, almost 104 economies still have laws preventing women from working in specific types of jobs; 59 economies have no laws on sexual harassment in the workplace; and in 18 economies, husbands can legally prevent their wives from working (HTTP10; Whitmore and Nunn 2017). These legal frictions downplay the contributions that almost half of these nations' human resources could contribute toward their economic development and global standing. Due to these legal and cultural reasons, the labor force participation rate for women in the 25–54 age group is 63% compared to 94% for men – an almost 2:3 ratio. When including younger labor force (aged 15 years and above) and older women (55 years and above), in 2018 women's global labor force participation rate was even lower, at 48.5%, 26.5% below that of men (HTTP10). These cause ripple effects on current and future employment status for women. Per United Nations, in 2017, the global unemployment rates for men and women stood at 5.5% and 6.2%, respectively (HTTP10), and it is projected to stay unchanged in the coming few years. The globalization and post-Fordist era have forced more women into unemployment and informal economies (HTTP9; HTTP10; Sharma 2017). The latest available data shows that the share of women in informal employment in developing countries is 4.6% more than that for men, when including agricultural workers, and 7.8% more when excluding them (HTTP10). In India alone, the overall share of labor in informal industries is enormously high, more than 75% of total labor, and women comprise the largest shares in the most vulnerable of employments (Sharma 2017).

Women are paid less than men and the global gender wage gap is estimated to be 23%. Women bear disproportionate responsibility for unpaid care and domestic work that are essential to the functioning of the economy, but often goes uncounted and unrecognized and the long-term opportunity costs are very high for such households where a talented and skilled woman stays home for childcare due to cultural reasons (HTTP10). A delay in entering into the labor market further jeopardizes their long-term competitiveness and growth prospects (Coe et al. 2012). Gender inequalities in employment and job quality result in gender gaps in access to social protection acquired through employment, such as pensions, unemployment benefits, or maternity protection. Globally, an estimated 40% of women in waged employment do not have access to social protection or access to financial institutions or bank accounts (HTTP11). While 65% of men report having an account at a formal financial institution, only 58% of women do worldwide (HTTP12). In addition, the digital divide all across the world is also quite gendered, with a majority (3.9 billion) of total population being offline living in rural areas, being less educated, and being women/girls (HTTP9; HTTP13). Thus, women are less likely to be entrepreneurs and face more disadvantages when starting businesses, which further constrain their being in high leadership positions (HTTP1; HTTP14). As per the United Nations, only 5% of the Fortune 500 CEOs are women (Cheryl 2014). The economic costs, which are reflections of the human and social costs to the global economy of discriminatory social institutions and violence against women, is estimated to be approximately USD 12 trillion annually (McKinsey & Company 2015). In short, it is no surprise that women's work has been deemed "invisible," "insignificant," and hence "invaluable" (Coe et al. 2012; HTTP9). Such differentiation of gendered work – largely understood as division of labor – is nothing new, but its economic repercussions in contemporary times are enormous as a large share of low- and middle-class women's professional activities are decided by the free time they can afford toward performing perceived economically valued professional activities outside of home – i.e., in the public spaces.

Theoretical Contexts

Scholars have used various theories and frameworks to explain the over/under-representation of women in specific types of industries. The *human capital theory* (HCT) assumes that gender concentrations are driven by the oversupply of labor with specific skill types which eventually land them those positions. Thus, the oversupply of the gender (or race) is the reason for having more women in specific industries compared to men, because far greater number of women apply for those positions than men do (HTTP1; HTTP9). As such, the fact that more women engage in service-sector/teaching professions as against more males in banking/financial or other professional jobs (Coe et al. 2012) is purely an outcome of demand and supply. HCT suggests that this occurs due to an oversupply of men or women for a particular kind of work, and hence sorting takes place outside of the labor market, i.e., it is by choice that men or women choose specific types of jobs, and that the labor market should not be blamed for over/under-representation of men versus women across various industries.

The *neoclassical economics*, alternatively, suggests that men and women looking for jobs are divorced from their overall social contexts, and search with a purely economic motive in mind. Thus, preferences, prejudice, discrimination, expectations, hopes, insecurities, etc., are seen as exogenous to labor markets. Though this theory suggests one to divorce job types from the socio-economic contexts of a person, in the real world, though, this rarely happens and socio-economic and political contexts of people do affect the types of jobs and opportunities that become available to them. In addition, if neoclassical economic theory were to hold, why would one notice over-concentration of people of specific races/ethnicities or gender in specific types of jobs? Finally, what about the household responsibilities that inherently is assumed to be a woman's responsibilities? What about the lack of information sharing and network regarding labor market opportunities that are accessible to those enjoying the public spaces? What about withholding of information from certain gender (and race) that deprives them of good opportunities? These are questions that cannot be fully answered by simply one or the other theory, and hence responding to the most contemporary spatial patterns of gender by industry may help identify the intersectionalities of these theories and other intervening constraints.

Lastly, the *labor market segmentation theory* rejects the view that job allocation is an outcome of skills, qualifications, and experience of applicants. Instead, it argues that each segment has its own distinctive set of rules that govern access to it, and mobility within it (Coe et al. 2012). Thus, the historical, institutional, and geographical contexts of a labor market are far more important than the human capital endowments of individuals. Early versions of segmentation theory identified primary and secondary labor markets, with the former characterized by stable, well-paid jobs generally occupied by white males, and the latter comprising unstable, low-paid jobs with few prospects for internal promotion. Recent versions of segmentation theory go beyond this dual labor market. The identities of employees, the jobs that they occupy, and the contexts provided by families, schools, neighborhoods, etc., are socially reproduced. In sum, then, the labor market segmentation theory very succinctly points toward the rules and political economy within each institution that get manifested in various forms of segregated occupational/ labor patterns, and gender gap is one of these manifestations produced within the larger premise of these institutions.

To sum up, existing literature points toward gender pay gaps and occupational segregation in the labor market due to a variety of reasons, and while some progress has been made due to a greater participation of women in a few male-dominated industries, the actual change and progress is not yet known. This chapter analyzes the degrees of male-versus-female presence in major industry types and changes during 2000–2017. It will also examine the spatial disparity in the over- and under-representation of gender in all major industry types in 2017 and situate its patterns through one of the existing three theories.

Research Design

This analysis focuses on the counties as a scale of analysis for the entire United States, excluding Puerto Rico.

Five years' ACS estimates data for 2013–2017 and the 2000 data from the Census of U.S. Bureau are used for conducting basic demographic analyses (figure 4.1) as well as for computing the location quotients (LQ) by gender

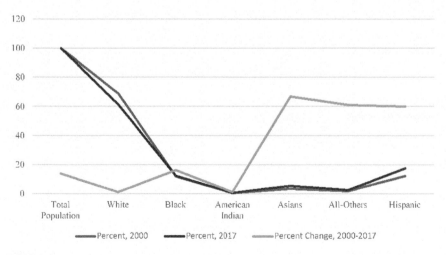

FIGURE 4.1
Demographic profile of study area (3,136 counties of the United States, excluding Puerto Rico).

for industry types for both 2000 and 2017. First, the descriptive statistics provide an overview of gender participation by industry across all major industry types in both years (figure 4.2 – top and middle); thereafter the degrees of change are analyzed across all industries (figure 4.2 – bottom). This is followed by computation of LQ values by gender by industry types for both years. Thereafter, Pearson's bivariate correlations analyses flesh out the degrees of association between human capital variables proxied by educational attainments and the computed LQs by industries. Later, ArcGIS is used to create maps for the computed LQ values for 2017. This helps understand gender-based spatial disparity in over/under-representation in major industry types (figures 4.3-A through 4.3-C). LQ values are computed using the specifications in Moineddin et al. (2003) and Wong (2003), among others. A generic equation for location quotient (*LQi*) for an industry *i* in a county can be written as:

$$LQi = \left(\left(ei / e \right) \right) \left(\left(Ei / E \right) \right) \tag{1}$$

where, *ei* is the employment in sector *i* in the local region/county; *e* is the total employment in the country; *Ei* is the employment of the gender (male/female) in industry *i* in the national economy, and *E* is total employment in the national economy (adding all types of industries taken together). After computing the LQs for employment in major industries across the United States, a cartographic analysis of computed values of LQ is conducted to gain perspectives on their spatial variation across the genders. This provides important insights concerning which parts/sub-regions of the United States have higher or lower representation of women versus men in specific industry types. The LQ maps could point toward space- and location-based dynamism that could be instrumental in creating and sustaining gender-biased differences in specific industries/economies, which could be reflective of long-held historical, socio-economic, political, and cultural contexts of these spaces.

Analyses and Findings

Analysis of Study Area and Industry-Based Employment by Gender

This analysis focuses on 3,136 counties of the United States, excluding Puerto Rico and including Alaska and Hawaii Islands. In the process of matching the shapefiles, few counties had to be removed due to data inconsistencies. As illustrated in figure 4.1, the total population of the United States grew from 281.39 million (2000) to 320.92 million (2017), with an almost 14.05% change, whereas the share of Whites declined from 69.13% (2000) to 61.45%

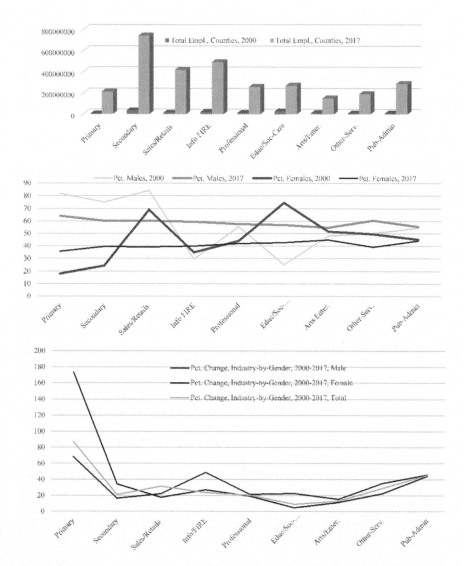

FIGURE 4.2

Total employment by industry, 2000 and 2017 (top); total employment by industry by gender, 2000 and 2017 (middle); and percent change in total employment and by gender (bottom) (civilians, 16 years/above) *Note*: Primary: Agriculture, forestry, fishing, hunting, mining; Secondary: Construction, manufacturing, transportation, warehousing, utilities; Sales/Retails: Wholesale/ retail-trade, Info/FIRE: Information/Finance, insurance, real estate; Professional: Professional, scientific, management, admin, waste management services; Educ/Soc-Care: Educational, healthcare, social services; Arts/Enter.: Arts, entertainment, recreation, accommodation, food services, Other-Serv.: Other services, no public admin; Pub-Admin: Public administration.

(2017). Most other racial/ethnic groups showed significant gains during this time period.

Figure 4.2 illustrates industry-based employments in 2000 and 2017 (top), percent share by industry by gender in 2000 and 2017 (middle), and percent change by industry for total labor and by gender (bottom). These three graphs are quite informative in that across all industry types, substantive gains in total employment have occurred for females too, and despite these positive changes, the percent participation by females still remains far lower than that for males. The industry-by-gender graph (figure 4.2 – bottom) also shows specific industries where significant gains have been made regarding female participation. There is an overall growth of 94% in the total numbers of labor employed during 2000–2017. It is no surprise that the share of females in the primary sectors (agriculture, forestry, fishing and hunting, and mining) is far lower than that of men in both years, even though the overall percentage gain in females in this sector has increased significantly by almost 173.62% (figure 4.2 – middle). However, the overall share in this sector, largely considered a masculine industry, is still dominated by men, who hold 81.87% and 64.09%, respectively, in 2000 and 2017. For secondary sectors (construction, manufacturing, transportation, warehousing, and utilities), the story stays the same, and as noted regarding the masculine versus feminine activities, this industry has male over-representation across both years (75.32% and 60%), and like primary sectors, women have increased their percentage share in this industry as well during 2000–2017. Likewise, figure 4.2 (middle) also suggests interesting divisions of labor along gender across various industry types. The industries where gender-based divisions seem relatively more equitably distributed include arts, entertainment, recreation, accommodation and food services, and public administration across both years, though the share of females has shifted gradually toward male presence in other services, excluding public administration in 2017; in 2000, the shares of both genders were quite equitable in this sector.

In terms of percent change across gender by industry (figure 4.2 – bottom), females have scored the most changes in the sectors of agriculture, forestry, fishing and hunting, and mining (173.62%), public administration (44.38%), information and FIRE (finance, insurance and real-estate) services (27.42%), and public administration (44.37%). For males, though, the largest scorers still include the primary sector (68.02), information/FIRE (48.58%), and public administration (46.24%). This simple analysis of industry-based employment shifts over time illustrates a macro-scale shift toward information/FIRE, along with other service sectors – a hallmark of American post-Fordism since the 1970s. These figures clearly indicate gendered division of labor, in that even though women's participation has grown in industries traditionally known for male dominance, women are still under-represented across these industries despite significant gains during 2000–2017 (Lexmond 2014).

Industry-Based Representation of Gender and the Theories of Gendered Segmentation

A Pearson's bivariate correlations analysis between human capital variables, proxied by educational attainments and location quotients for industry by gender, illustrates very interesting patterns. Population size in 2000 and 2017 as well as its change associate with lower representation for males in primary sectors in 2017, though these are positive and significant for women. Population size and its change have positive associations for males and females for wholesale/retail trade, information/FIRE services, and for professional, scientific, and management services. Negative associations exist between education/health care–related services for both genders, for arts and entertainment/recreation industry, for other services, and for construction/secondary industries for males in 2017. These generic trends suggest more about the larger macro-scale shift of industries in the United States that have tilted heavily toward tertiary-based economy – wherein all genders have significant and positive association (table 4.1).

Regarding educational attainment categories that represent human capital resources of a place/space, in general, males with no schooling/lower education (some college/associates/GED or below) have higher representation in primary, secondary, and other services such as entertainment/recreation and in public administration. Females have negative association with these sectors in general. However, females with slightly better education (some college/associates/GED), have positive association with primary sector. For those with bachelor's degrees, males have negative association with primary and secondary sectors and women have a positive association; likewise, it is positive for males and females both in the sectors of wholesale/retail trade, information and FIRE, professional/scientific, etc., and for females only in public administration. Males with professional degrees/master's/doctorates exhibit over-representation with wholesale-retail trade, information/FIRE, and professional/scientific in 2017; this also contributes to a higher representation of females in these same industries.

In contrast, when examining the association of female human capital versus industry over/under-representation, those with no high school diplomas have negative association with agriculture/mining; construction/manufacturing; and education and health care services. Females with high school diplomas have negative association with primary (agriculture/mining) sector, secondary sector (construction/transportation), and information/FIRE. However, presence of females with high school diplomas does attribute to a lower representation of males in wholesale/retailing, information/FIRE, education/health care–related and other services for males. When females with bachelor's degree are present, the LQs for females become positive in primary and secondary sectors, as well as in wholesale/retail-trade, information/FIRE, and in professional/scientific sectors. However, it stays lower in educational/healthcare services and in entertainment/recreation industries.

TABLE 4.1

Bivariate Correlationships for Location Quotients for Employment by Industry by Gender, 2017

Education, Year	Primary		Secondary		Sales/Retails		Info/FIRE		Professional		Educ/Soc-Care		Arts/Enter.		Other-Serv.		Public Administration	
	M	F	M	F	M	F	M	F	M	F	M	F	M	F	M	F	M	F
POP, 2000	-.102**	.051**	-.085**	.123**	.104**	.104**	.104**	.160**	.036*	.041*	-.051**	-.080**	-.030	-.025	-.087**	.023	.000	-.021
POP, 2017	-.103**	.056**	-.087**	.127**	.106**	.109**	.106**	.164**	.035	.041*	-.052**	-.083**	-.031	-.026	-.089**	.022	-.001	-.024
Change, 2017-2000	-.049*	.005	-.025	.067*	.059**	.044*	.059**	.075**	.031	.027	-.022	-.058**	-.002	-.007	-.034	.029	-.010	-.041*
Males, 2000	M	F	M	F	M	F	M	F	M	F	M	F	M	F	M	F	M	F
No-Schooling	.217**	-.043*	-.021	-.138**	-.131**	-.087**	-.131**	-.172**	-.057**	-.026	.078**	.123**	.054**	.012	.062**	-.077**	.061**	.028
No-HS-Diploma	.153**	-.170**	.091**	-.145**	-.157**	-.165**	-.157**	-.226**	-.010	.004	.059**	.117**	.065**	.111**	.151**	-.031	.038*	.053**
HS-Diploma	.029	-.039*	.098**	-.023	-.065**	-.074**	-.065**	-.071**	-.025	-.029	-.011	.015	-.006	.000	.094**	.008	-.011	.002
Some-Coll/GED/Assoc.	-.035*	.187**	-.057*	.015	.017	.067*	.017	.105**	-.072**	-.055**	-.012	-.028	-.027	-.031	-.117**	-.040*	.017	.031
Bachelor's	-.109**	.136**	-.120**	.075*	.127**	.141**	.127**	.172**	.057**	.032	-.021	-.052**	-.027	-.055*	-.143**	-.005	-.058**	-.007
Prof./MS/Doct.	-.134**	.043*	-.164**	.129**	.151**	.128**	.151**	.200**	.089**	.069**	-.032	-.093**	.014	-.039*	-.128**	.057*	-.038*	-.030
Females, 2000	M	F	M	F	M	F	M	F	M	F	M	F	M	F	M	F	M	F
No-Schooling	.153**	-.170**	.091**	-.145**	-.157**	-.165**	-.157**	-.226**	-.010	.004	.059**	.117**	.065*	.111**	.151**	-.031	.038*	.053**
No-HS-Diploma	.120**	-.186**	.079**	-.084**	-.110**	-.133**	-.110**	-.188**	-.010	.015	.053**	.076**	.045*	.060*	.143**	-.008	.037	-.006
HS-Diploma	-.036*	.188**	.097**	.108**	.046*	-.005	.046*	.034	.023	-.021	-.069**	-.109**	-.023	-.034	.045*	.063**	-.004	-.086**
Some-Coll/GED/Assoc.	-.074**	.188**	-.029	.005	.054**	.088**	.054**	.087**	-.065**	-.049**	.009	-.007	-.068**	-.075*	-.091**	-.035	.018	.014
Bachelor's	-.100**	.151**	-.095**	.038*	.114**	.139**	.114**	.140**	.033	.031	-.027	-.030	-.042*	-.014	-.146**	.008	-.065**	.022
Prof./MS/Doct.	-.183**	-.009	-.169**	.163**	.154**	.150**	.154**	.211**	.122**	.085**	-.053**	-.087**	.012	-.035	-.129**	.086**	-.055**	-.041*

(Continued)

TABLE 4.1 (CONTINUED)

Bivariate Correlationships for Location Quotients for Employment by Industry by Gender, 2017

Education, Year	Primary M	Primary F	Secondary M	Secondary F	Sales/Retails M	Sales/Retails F	Info/FIRE M	Info/FIRE F	Professional M	Professional F	Educ/Soc-Care M	Educ/Soc-Care F	Arts/Enter. M	Arts/Enter. F	Other-Serv. M	Other-Serv. F	Public Administration M	Public Administration F
Males, 2017																		
No-Schooling	0.025	-.048**	0.011	-.020	0.004	-0.029	0.004	-.044*	0.006	0.034	0.000	.064**	0.010	-0.025	0.002	-.036*	.062**	-0.029
No-HS-Diploma	.208**	-.107**	.047**	-.161**	-.170**	-.153**	-.170**	-.218**	-.046**	-0.005	.063**	.102**	.070**	.047*	.126**	-.044*	.054**	.051**
HS-Diploma	.107**	-.064**	.116**	-.075**	-.150*	-.121**	-.150**	-.150**	-.059**	-0.030	-0.009	.062**	0.027	0.037	.127**	-0.013	0.010	.049*
Some-Coll/ GED/ Assoc.	0.008	.147**	0.028	-.057**	-0.016	0.007	-0.016	0.007	-.082**	-.076**	0.023	-0.002	-.052**	-0.032	-.048**	-.057**	0.033	0.016
Bachelor's	-.134**	.126**	-.101**	.063**	.141**	.148*	.141**	.169**	.055**	0.031	-0.017	-.052**	-.037*	-.040*	-.140**	-0.012	-.063**	0.002
Prof./MS/ Doct.	-.184**	0.028	-.173**	.151**	.174**	.129**	.174**	.213**	.123**	.084**	-0.031	-.117**	-0.008	-0.015	-.137**	.068**	-.056**	-0.027
Females, 2017																		
No-Schooling	0.016	-.011	-0.032	0.006	0.011	-0.006	0.011	-0.013	-0.008	.046**	-0.002	0.030	0.010	0.006	-0.007	-0.001	.036*	-.045*
No-HS- Diploma	.205**	-.110**	.036*	-.113**	-.115**	-.108**	-.115**	-.176**	-.044*	-0.011	.065**	.076**	0.031	0.028	.109**	-0.024	.057**	0.011
HS-Diploma	0.030	-.106**	.107**	.061**	-.035*	-.073**	-.035*	-.037	-0.007	-.036*	-.055**	-0.017	0.005	0.019	.087**	0.024	-0.010	-.059**
Some-Coll/ GED/ Assoc.	-.080**	.059**	.040*	0.010	0.034	0.032	0.034	0.026	-0.035	-.042	.038*	-0.018	-.037	-0.009	-0.030	0.005	.038*	-0.010
Bachelor's	-.159**	.128**	-.102**	.073**	.159**	.165**	.159**	.173**	.051**	.039*	-0.026	-.061**	-0.033	-.048**	-.136**	0.020	-.063**	-.025
Prof./MS/ Doct.	-.199**	-0.011	-.162**	.182**	.173**	.152**	.173**	.232**	.148**	.103**	-.071**	-.081**	0.001	-0.025	-.127**	.083**	-.052**	-.031

* Correlation is significant at the 0.05 level (P<0.05) and ** Correlation is significant at the 0.01 level (P< 0.01)

With greater shares of females with professionals/master's/doctorates, the r-values for females are positive in secondary sectors, wholesale/retail trade, information and FIRE, professional/scientific/management, and in entertainment and other services, whereas it is negative for education/health care services. These suggest an interesting phenomenon of gradual shifts in gendered divisions of labor, with gaps narrowing across specific industries.

Finally, descriptive statistics of computed values of location quotients (LQs) for both genders by industry types (table 4.2) illustrate interesting things about U.S. labor markets. Males have substantial over-representation in professional sectors (LQ = 5.7), education/health care (LQ maximum value of 3.16), arts/entertainment (LQ maximum value of 5.78), other services (LQ maximum value of 5.18), and public administration (LQ maximum value of 12.4). This is also illustrated in the location quotient maps (figure 4.3). Interestingly, while the share of women has increased in many industries, compared to men they still exhibit significant lower presence, with

TABLE 4.2

Descriptive Statistics (Percentiles) for Location Quotients (LQ) by Industry by Gender, 2017

Industry Sectors	LQ, Males, 2017					LQ, Females, 2017				
	Min.	Max.	35th	50th	65th	Min.	Max.	35th	50th	65th
Agriculture, forestry, fishing, hunting, mining	0.00	1.18	0.05	0.05	0.06	0.00	0.26	0.02	0.03	0.04
Construction, manufacturing, transportation, warehousing, utilities	0.00	0.10	0.03	0.03	0.03	0.00	0.10	0.03	0.04	0.04
Wholesale and retail trade	0.00	0.18	0.05	0.06	0.07	0.00	0.15	0.03	0.03	0.04
Information, finance, insurance, real estate	0.00	0.16	0.05	0.05	0.06	0.00	0.09	0.03	0.04	0.04
Professional, scientific, management Admin, waste management services	0.00	5.71	0.57	0.69	0.83	0.00	0.29	0.03	0.03	0.04
Educational, health care, social services	0.00	3.16	0.17	0.19	0.20	0.00	0.33	0.03	0.03	0.04
Arts, entertainment, recreation, accommodation, food services	0.00	5.78	0.28	0.31	0.35	0.00	0.53	0.03	0.04	0.04
Other services (excluding public administration)	0.00	5.19	0.24	0.27	0.31	0.00	0.32	0.03	0.03	0.04
Public administration	0.00	12.41	0.48	0.53	0.58	0.00	0.31	0.03	0.04	0.04

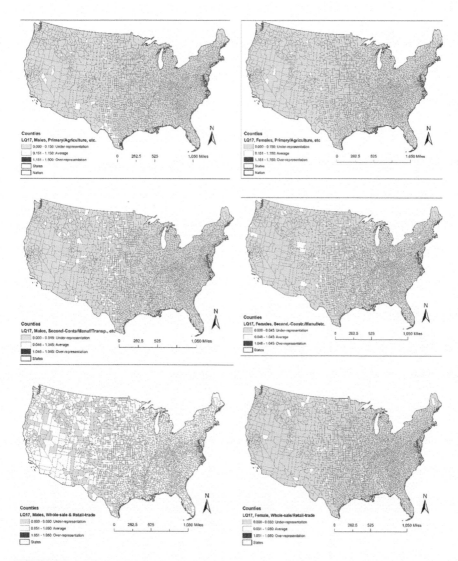

FIGURE 4.3
A. Location quotients for primary (top), secondary (middle), and wholesale/trade (bottom).

their LQ values very low across all industries. The maximum LQ value for females are 0.529 (arts/entertainment), 0.327 (education), 0.318 (other services), 0.310 (public administration), and 0.289 (professional) sectors. These values are alarmingly low compared to those of their male counterparts, who score above 3.0 across all major industries. This speaks volumes about gender-based division of labor and the lack of female participation across almost all industry types, even though data suggests positive growth during

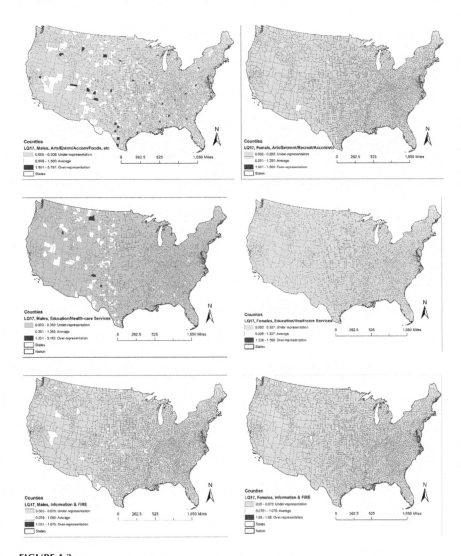

FIGURE 4.3
B. Location quotients for arts/entertainment (top), education/health (middle), and information/FIRE (bottom).

2000–2017, also confirmed by the United Nations report and the McKinsey & Company report of 2017 (Krivkovich et al. 2017).

Spatial Analyses of Industry-Based Employment by Gender

Using GIS maps as a visual tool, the location quotients maps for various industries for both genders for 2017 suggest interesting patterns of gendered division of labor across U.S. counties (figures 4.3 A, B, and C). These location

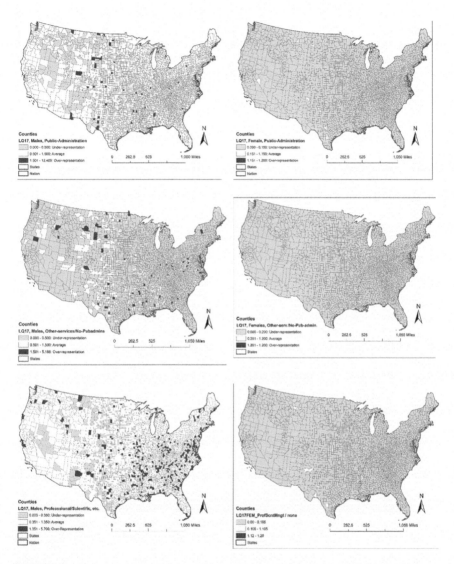

FIGURE 4.3

C. Location quotients for public administration (top), other services (middle), and professional/scientific (bottom).

quotient maps for industries illustrate space-/county-based under- and over-representation by gender and are hence reflective of overall presence (or the lack thereof) of human capital resources or the neoclassical economics or the market labor segmentation theories that create gender-based spatial patterns in each industry type. In yet another analysis, Sharma (2016) found women in the American southeast as more suited for feminine jobs performed within private spaces, whereas women in the northeast or the Pacific West Coast found these cultural nuances of the American southeast very archaic (see Sharma 2016 for more such illustrations). Such space-/region-based nuances are critical for understanding and creating space-/county-based policies that can best address their economic and human resource development needs (proxied by education in this analyses) which could be critical to the labor market. Likewise, random patches of dark-shaded counties in specific industries represent males' over-representation, and this needs further analysis. They may provide insights into the intersectionalities between local cultures, policies, and historical contexts, and if educational attainments and skill differences could be the reasons for gendered-based differences in these industries.

Conclusions and Policy Implications

This analysis was conducted with an objective of identifying the industries where gender-based over-/under-representation may persist. This could be critical for creating an analytical framework for applying gender and economic development concepts in place-based policymaking. In any household, gender structures and gender roles are defined by the tasks conducted by men and women, and scholars have suggested strong linkages between gendered labor and location-specific cultural nuances. Despite these spatial differences, it holds true that time and again women end up performing the larger share of unpaid and invisible household chores as these are performed inside private/ domestic spaces, which make them undervalued, unappreciated, and invisible (Coe et al. 2012). These affect the conditions of financial provisioning, which eventually limit women's empowerment since private spaces strictly constrain the professional linkages and networks critical for professional growth of women (Coe et al. 2012). This analysis finds that across the board, women have far lower representation than men in all sectors of industries. Compared to men, a very small share of women work, and, despite positive gains in all industry types during 2000–2017, the dominance of men in all industries continues. This, however, could be deceptive since women's overall participation has stayed lower compared to men across all industries in U.S. counties for centuries, since the concept of "industry" has been associated with the tasks that comprise paid work only. However, if this analysis was conducted using only women as the universal population, and their over-/under-representation across industry types during this time period, that would indeed flesh out more insightful results concerning the actual change and continuity about gendered economy. That could be useful in incentivizing the industries that currently rank low regarding women's representation, which could also be critical for potential planners and policymakers.

This analysis focused on gender participation in industry types and its change and continuity over time. The findings can be useful in creating a gender-inclusive work environment (Landau and Lewis 2018; Lexmond 2014). First, there is a pressing need to do more, and even though most organizations realize these needs based on a recent reporting by McKinsey & Company (2015) and Krivkovich et al. (2017), not much is being done at the levels of institutions and households – the very scale at which gender discrimination begins. It is encouraging news that many companies and industries have now started showing their commitment toward gender diversity and inclusion, which is at an all-time high for the third year in a row (McKinsey & Company 2015; Krivkovich et al. 2017; Lexmond 2014). However, there is still a lot more that can be done. There is tremendous scope for skill development and for developing location-specific human capital resources for women's fuller participation in construction, manufacturing, and other industries that are well-perceived and accepted based on the cultural nuances of places/spaces. In some contexts, these trainings and skills could be imparted to break the age-old inertias of gendered limitations such that an economic equilibrium could be achieved. These include location- and region-specific training of women and their inclusion in industries that have typically been dominated by men (see HTTP2; HTTP3 and Demirguc-Kunt et al. (2015) for how positive growth has occurred across these male-dominated industries and the presence of talent and skill among women have been capitalized on with necessary policies of inclusion from the government). As indicated by some scholars, one of the most powerful reasons for the lack of progress concerning women's inclusion in many industries is that we have missed out on some blind spots when it comes to diversity and inclusion, and owing to the fact that women comprise almost half of the country's population, we can't solve the problems that we don't see or understand clearly and completely (Krivkovich et al. 2017; Momani & Sumaya 2017). Some of these blind spots include lack of coordination and effective communication across the various wings within federal and state governments that are expected to commit toward gender diversity and inclusion. Unfortunately, many of these entities still work in a very bureaucratic, culturally insensitive, and old fashioned styles, which pose hurdles for women's effective inclusion in work spaces (author's observations from field work and professional career growth).

There is also a need to make changes in many state and national policies to help women become equal participants in economy. As indicated earlier, the most important reasons for the lack of females participation across formal economy is their struggle to cope up with child care and family care duties. A majority of U.S. work institutions does not provide paid maternity leave and/or subsidized child care and family care benefits. These inadvertently fall under the female domain regardless of the socio-economic status of women, and this limits their participation and fuller growth in many industries, despite their educational attainments and skills. If the federal and state governments make gender-sensitive care benefits mandatory, it could boost fuller and productive participation of women across industries (Whitmore and Nunn 2017). These can be very critical for a nation's holistic growth (see Reeves and Venator's

2014 analyses of their proposed reforms on expanding female opportunity in work spaces by providing for child care credits or the FAMILY Act's paid leave provisions). Even the United Nations has suggested numerous steps to boost gender participation across many industries at global scale. The UN suggests that by empowering women in economy, one could close the gender gaps in the world of work, which is critical for the holistic development of any society and country (HTTP10). By increasing women's educational attainment, one could contribute toward their economic empowerment and inclusive economic growth (Landau and Lewis 2018). As the proverb goes, "You educate a man, you educate an individual; you educate a woman, you educate a generation." Further, the legal restrictions on women from having similar opportunities as men also constrains the levels of investments in women's human capital resource development policies such as creating gender-sensitive policies (e.g., paid maternity benefits/leaves), family leaves, changing the political and cultural contexts by slightly lowering the bars for women's entry in specific industries that are typically male dominated, and the like.

Ingram and Nora (2018) also suggest that one of the very important reasons for significant gaps in gendered economy is the lack of good-quality data on gender (Lexmond 2014). As of now, gendered data is available in bits and pieces, scattered across various variables on different data sources, which makes it cumbersome, difficult, and time consuming to engage in holistic meaningful analysis. This needs policy and institutional intervention so that varieties of data pertaining to gender could be made available at point source. That would encourage scholars as well as policymakers to conduct suited research that can promote gender equity (Ingram and Nora 2018). These steps are critical for encouraging a whole new upcoming generation since these will help break the historical glass ceiling, which can be critical to women's participation in certain spheres of labor market. Policies also need to be made that can reduce gender gaps in earnings in similarly skilled industries/occupations. Since women are over-represented in informal and vulnerable occupations, it is equally important for the employers to create safe work spaces for them. One of the deterrents for women's participation in male-dominated industries has been the lack of voices and a safe work environment, which constrains women from working despite their education and skills. With positive steps toward creating safe work environments, women will participate in all industries as they are equally committed and skilled to create an economically healthy nation and productive citizens.

References

Cheryl. (2014). *Women CEOs of the S&P 500*. http://www.catalyst.org/knowledge/women-ceos-sp-500

Coe, Neil M., Kelly, Philip F., and Yeung, Henry W.C. (2012). *Economic Geography: A Contemporary*. Introduction by Neil M Coe, Philip F. Kelly, and Henry W.C. Yeung. 2nd Eds, Wiley, USA. ISBN-13: 978-0470943380

Demirguc-Kunt and others. (2015). *The Global Findex Database 2014: Measuring Financial Inclusion Around the World*. Policy Research Working Paper 7255. Washington, DC, World Bank. Last accessed on 7/22/2019 at http://documents.worldban k.org/curated/en/187761468179367706/pdf/WPS7255.pdf

Furuseth, O. J., and Smith, H. A. (2006). From Winn-Dixie to Tiendas: the remaking of the New South, in *Latinos in the New South: Transformations of the Place*, Eds Heather Smith and Owen Furuseth, New York, NY, Ashgate Publishers, 1–17.

HTTP1: Predescu, Sebastian. (2017). *Data Reveals How the Percentage of Women in Leadership Has Changed Over the Past 15 Years*. Last accessed on 8/1/2019 at https ://business.linkedin.com/talent-solutions/blog/trends-and-research/2017/ data-reveals-how-the-percentage-of-women-in-leadership-has-changed-over-t he-past-15-years

HTTP2: Women in Industry Service. Last accessed on 08/06/2019 at https://ww w.encyclopedia.com/history/encyclopedias-almanacs-transcripts-and-maps/ women-industry-service

HTTP3: Women in Manufacturing: Stepping up to Make an Impact that Matters, A Report Prepared by The Deloitte and The Manufacturing Institute. Last accessed on 08/06/2019 at: https://www.census.gov/newsroom/blogs/ra ndom-samplings/2017/10/women-manufacturing.html

HTTP4: Quick Take: Women in Male-Dominated Industries and Occupations (2018). Last accessed on 08/05/2019 at https://www.catalyst.org/research/women-in-male-dominated-industries-and-occupations/

HTTP5: ILO Estimate about Employment in Industry. Last accessed on 08/06/2019 at https://data.worldbank.org/indicator/SL.IND.EMPL.FE.ZS?view=chart

HTTP6: ILO, Women at Work: Trends 2016 (Geneva, 2016). Last accessed on 8/6/2019 at http://www.ilo.org/wcmsp5/groups/public/---dgreports/---dcomm/---publ/documents/publication/wcms_457317.pdf

HTTP7: Employed Persons by Detailed Industry, Sex, Race, and Hispanic or Latino Ethnicity; Labor Force Statistics from the Current Population Survey, Bureau of Labor Statistics. Last accessed on 08/06/2019 at https://www.bls.gov/cps/ cpsaat18. html

HTTP8: Degrees Conferred by Postsecondary Institutions, by Level of Degree and Sex of Student: Selected years, 1869–70 through 2025-26, report of National Center for Education Statistics. Last accessed on 08/06/2019 at https://nces. ed.gov/programs/digest/ d15/tables/dt15_318.10.asp

HTTP9: Why Gender Matters in Economy? Last accessed on 8/5/2019 at https://ww w.arts.ubc.ca/why-gender-matters-in-economics/

HTTP10: UN Economic Development. Last accessed on 8/5/2019 at http://www .unwomen.org/en/what-we-do/economic-empowerment/facts-and-figures

HTTP11: Global Entrepreneurship Monitor. (2017). *GEM 2016/2017 Women's Entrepreneurship Report*. *Women's Entrepreneurship Report*. https://www.gem consortium.org/report/49860

Ingram, George, and O'Connell, Nora. (2018). *A Leap in Gender Equality Begins with Better Data*. Last accessed 5/4/2018 at https://www.brookings.edu/blog/futu re-development/2018/05/04/a-leap-in-gender-equality-begins-with-better -data/

Krivkovich, Alexis, Robinson, Kelsey, Starikova, Irina, Valentino, Rachel, and Lareina, Yee. (2017). *Women in the Work Place 2017*. Report by McKinsey & Company, 1–6.

Landau, Kelsey Lewis. (2018). *Toward Gender Equality and Inclusivity in Oil, Gas, and Mining*. Last accessed on 7/3/2018 at https://www.brookings.edu/blog/fixg ov/2019/03/29/toward-gender-equality-and-inclusivity-in-oil-gas-and-mi ning/

Lester, W. T., and Nguyen, M. T. (2015). The economic integration of immigrants and regional resilience, *Journal of Urban Affairs*, 38(1): 42–60.

Lexmond, Jen. (2014). Gendered character. *Brookings Report*. Last accessed on 5/1/2019.

Lichter, Dan T., and Johnson, K. M. 2009. Immigrant gateways and Hispanic migration to new destinations, *International Migration Review*, 43(3): 496–518.

McKinsey & Company. (2015), *The Power of Parity: How Advancing Women's Equality can Add $12 Trillion to Global Growth*. Last accessed on 7/20/2019 at https://ww w.mckinsey.com/~/media/McKinsey/Featured%20Insights/Employment%20 and%20Growth/How%20advancing%20womens%20equality%20can%20add %2012%20trillion%20to%20global%20growth/MGI%20Power%20of%20parity _Full%20report_September%202015.ashx

Mohl, R. (2007). *Hispanics and Latinos: Encyclopedia of Alabama*. Last acceded on 02/03/2016 at http://www.encyclopediaofalabama.org/article/h-1096?printabl e=true

Moineddin, R., Beyene, J., and Boyle, E. (2003). On the location quotient confidence interval, *Geographical Analysis*, 35(3): 249–256.

Momani, Bessma, and Attia, Sumaya. (2017). *Struggling Economy? Gender Equality can Fix That*. Last accessed on 1/10/2017 at https://www.brookings.edu/opinions/ struggling-economy-gender-equality-can-fix-that/

Robinson, Gerard. (2018). *A Tale of two Disparity Gaps*. Last accessed on 7/3/2018 at https://www.brookings.edu/blog/brown-center-chalkboard/2016/06/30/a-ta le-of-two-disparity-gaps/

Reeves, Richard V., and Venator, Joanna. (2014). Policies to expand women's opportunity. *Brookings Report*.

Sawhill, Isabel, V. (2016). The gender pay gap: to equality and beyond. *Social Mobility Memos. Brookings Report*.

Schneebaum, Alyssa. (2016). Illuminating the Role of Gender in the Economy in | *Wirtschaft neu denken: Blinde Flecken in der Lehrbuchökonomie*. Last accessed on 7/19/2019 at http://fgw-nrw.de/fileadmin/user_upload/Blinde_Flecken_der_L ehrbuchoekonomie_klein.pdf

Schuele, Finn, and Sheiner, Louise. (2018). *Hutchins Roundup: Slack and Inflation, the Gender Pay Gap in the Gig Economy, and More*. Last accessed on 5/2019 at https ://www.brookings.edu/blog/up-front/2018/06/28/hutchins-roundup-slack-a nd-inflation-the-gender-pay-gap-in-the-gig-economy-and-more/

Sharma, Madhuri. (2016). Spatial perspectives on diversity and economic growth in Alabama, 1990–2011. *Southeastern Geographer*, 56(3): 329–354. http://muse.jhu. edu/article/633139/pdf

Sharma, Madhuri. (2017). Quality of life of labor engaged in informal economy in the national capital territory of Delhi, India. *Khoj: The International Peer Reviewed Journal of Geography*, 4: 14–25. doi:10.5958/2455-6963.2017.00002.9. [Print-ISSN: 2350-1359; Online-ISSN: 2455-6963; *Indexed/Abstract with: Summon Proquest, Ebsco Discovery, CNKI Scholar, Google Scholar*].

Sharma, Madhuri. (2018). Economic growth potentials and race/ethnicity in Tennessee: diversity and economy, *International Journal of Applied Geospatial Research*, 9(2): 33–54.

Whitmore, Diane Schanzenbach, and Nunn, Ryan. (2017). The 51%: driving growth through women's economic participation. *Brookings Report*.

Wong, David W.S. (2003). Spatial decomposition of segregation indices: a framework toward measuring segregation at multiple levels. *Geographical Analysis*, 35(3): 179–194.

5

Social and Environmental Injustice Experienced by Female Migrant Workers in China: The Case of Guiyu Town

Ye Zhang

CONTENTS

Introduction

After the "Reform and Opening Up" policy was implemented in China in 1978, millions of rural farmers have migrated from their hometowns to urban cities to earn money. This migration is a result of the combined effects of the demand for cheap labor in urban areas – as a result of attracting business and investment – and farmers' break away from the planned economy and *huji zhidu* (governmental system of household registration).[1] This group of people – usually called "migrant workers (*nongmingong*)" – has attracted considerable attention from scholars across various disciplines. The feminization of labor (Ong 1991), which explains how female labor in developing countries is drawn into the "global assembly line," often features Chinese manufacturing industries. Figure 5.1 shows the number of female migrants in 31 provincial level administrative areas, exclusive of Hong Kong, Macao, and Taiwan. Various aspects of female migrant workers' vulnerable and marginalized status have also been discussed widely (e.g., Li 2004; Jacka 2014).

Female Migrants at Each Province

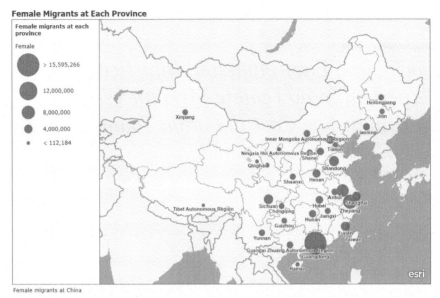

Female migrants at China

Esri, USGS | Esri, HERE, Garmin, FAO, NOAA, USGS | Esri, HERE, Garmin, FAO, NOAA, USGS

FIGURE 5.1
Map of the number of female migrants in 31 provincial level administrative areas (data come from the 2010 Population Census of the People's Republic of China [Census 2010]. The map is created using ArcGIS Online by Esri. ArcGIS® and ArcMap™ are the intellectual property of Esri and are used herein under license. Copyright © Esri. All rights reserved).

The rapid urbanization and economic development of China in the last three decades has been achieved to a significant extent at the expense of the environment and people's health. It has been argued that migrant workers in particular carry the burden of increased environmental pollution in both their places of work and their residence (Li and Liu 2006; Wang 2008; Ma 2010; Chen 2011; Chen et al. 2013; Sun and Shi 2016; Schoolman and Ma 2012). However, existing quantitative research on the environmentally vulnerable status of migrant workers widely ignores the factor of gender. Gender is either not considered at all or held to be an irrelevant variable with respect to environmental inequality (e.g., Nie 2013). On the other hand, even though qualitative studies (e.g., Jacka 2014; Pun 2005; Hanser 2008) have widely examined the intersected systems of oppression – based mainly on class, gender, and migrant status – faced by female migrant workers through illustrating their lived experiences and narratives, their environmentally vulnerable status has not received sufficient attention or been fully examined. This further reflects the gap between the quantitative, geographic approach and the qualitative, experiential approach in the discussion of social and environmental injustice. As this chapter shall show, GIS applications could be used as effective tools to fill this gap and make the discussion on gendered environmental injustice more solid and visualized.

This chapter shall focus on examining the environmentally vulnerable status of female migrant workers in China, using Guiyu town at Guangdong Province as an example. Guiyu town started importing electronic and electric waste (e-waste) from foreign countries and participating in the recycling business since the early 1990s. Under the development of global capitalism and the domestic transformation from a socialist to post-socialist/neoliberal society, Guiyu town has become one of the biggest e-waste recycling sites in China. However, the operating model of family workshop and the rough and brutal ways of dismantling e-waste that leave both nature and people directly exposed to environmental hazards has also brought with it serious environmental problems, including soil, water, and air pollution. This increase in serious pollution has also led to rigorous environmental governance from both central and local governments in recent years. A lot of (illegal) e-waste workshops have been banned, and local environmental supervision has become more and more strict.

The development of Guiyu town is inseparable from migrant workers, a large proportion of whom are women. Along with the migration tide, they come from other, poorer provinces, including Hunan, Sichuan, Anhui, Hubei, and Guangxi, among others, and make up most of the labor market of Guiyu town.[2] Using a qualitative GIS approach (Teixeira 2018), this chapter shall examine the environmentally vulnerable status of female migrant workers in the context of both the environmental pollution and governance of Guiyu town.

Research Methods

The plural understanding about environmental justice as distribution, recognition, participation, and capability (Schlosberg 2009; Walker 2012) directs us away from relying solely on quantitative research methods toward qualitative and experiential research methods. While the former methods are based mainly on a distributive sense of justice, the latter are based on other conceptions of justice and give more weight to the individual experiences, everyday lives, and narratives of people who experience environmental injustice. The qualitative research approach thus relies on employing research methods such as structured and unstructured interviews, participant observation, and participatory activist research (e.g., Krauss 1993; Tschakert 2009; Hope Alkon 2011). This project avoids putting different research methods into conflict with one another or favoring one over the other. Rather, it shall employ a qualitative GIS approach that integrates qualitative data collection and analysis with spatial analysis using GIS applications.

The qualitative data presented in this chapter are drawn mainly from the author's ethnographic fieldwork in Guiyu town from March to August 2018, which comprised deep participant observation, unstructured interviews, as well as informal dialogues and interactions with various groups of people. Ethnographic research allowed the author to immerse herself into the everyday lives of Guiyu residents to see how these different groups of people – mainly differentiated by class, migrant status, gender, and age – are differently influenced by the local environmental pollution. In addition, this chapter also uses data from other resources such as online stories, media reports, and scientific reports from NGOs and researchers. These resources are necessary because many historical moments can only be understood from previous investigations, news, and reports by mainstream and private media. Personal online story-telling, discussions, and comments in the local online community (e.g. *Guiyu Tieba*) also comprise important supplementary materials, helping this research to reflect more information and more diverse voices.

During the fieldwork, the author engaged with a total of 47 interviewees who indicated their consent to be participants after understanding the research aims and procedures. Tables 5.1 and 5.2 illustrate the composition of the participants and their home provinces, respectively. Figure 5.2 shows the map of their home provinces. These participants are quite diverse with respect to their social categories and previous/current occupations. The participants include both local and migrant people from several different provinces, female and male, and people at different ages. In terms of occupation, they include workers in e-waste workshops, hotel security personnel and cleaners who used to work in e-waste workshops, housewives undertaking odd jobs at home, local family workshop owners, school masters, town-level governmental officials, and village committees. The basic types of e-waste recycling work include: e-waste dismantling, sorting, and smashing different kinds of plastics, melting circuit boards, and other e-waste, separating precious metals through strong acid dissolution, etc. Due to the relationships the author has built over time, she was often invited to visit migrant workers' h omes to observe and engage in their everyday lives. It is through these activities and conversations that she has gained a fuller understanding of their lifestyles, living environments, experiences, stories, and perspectives.

TABLE 5.1

Composition of Participants

Subject populations	Gender	Number of participants
Migrant people	Male	12
	Female	18
Public officials[a]	Male	10
	Female	1
Local people[b]	Male	4
	Female	2
	Total number of participants	47

[a] Public officials include town-level government and village committee members.
[b] All public officials are local people.

TABLE 5.2

Participants' Home Provinces

Subject populations	Home province	Number of participants
Migrant workers	Guangxi Province	1
	Anhui Province	3
	Sichuan Province	15
	Hunan Province	4
	Jiangxi Province	1
	Hubei Province	4
	Zhejiang Province	1
	Henan Province	1
Local people	Guangdong Province	17
	Total number of participants	47

To visualize the social and environmental injustice experienced by female migrant workers, the author has also used the applications of ArcGIS Online and ArcGIS StoryMaps to spatialize the qualitative data of population census, photos, participants' information and narratives, observations in the walk-along tours, the author's field notes, etc. Figure 5.3 shows how ArcGIS StoryMaps has been used by the author in the processes of data collection,

FIGURE 5.2

Map of participants' home provinces (created using ArcGIS Online by Esri. ArcGIS® and ArcMap™ are the intellectual property of Esri and are used herein under license. Copyright © Esri. All rights reserved).

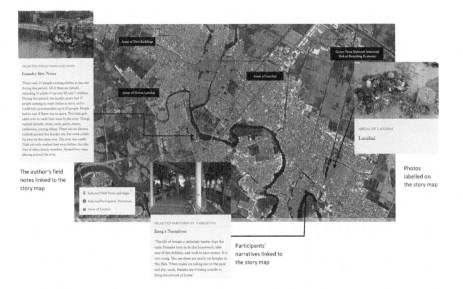

FIGURE 5.3
Visualization of qualitative data using ArcGIS StoryMaps (the screenshots in this figure are produced using ArcGIS StoryMaps by Esri, see https://storymaps.arcgis.com/stories/a253ea9a df0e48baa691c7b6e382cd07. ArcGIS® and ArcMap™ are the intellectual property of Esri and are used herein under license. Copyright © Esri. All rights reserved).

analysis, presentation, and dissemination. The combination of spatial and qualitative analysis can allow us to have a more contextual understanding of the social and environmental injustice against female migrant workers as lived experiences instead of just variables and regression models.

Environmental Injustice Experienced by Female Migrant Workers

Compared with local people, migrant people are more vulnerable to the effects of local environmental pollution and degradation. This is not only because of the harsh working conditions associated with the rough process of dismantling e-waste but also because of their living environments and lifestyles. The local environmental governance not only completely fails to consider them but also puts them in a more vulnerable position.

Everyday Lives of Female Migrant Workers

The general living conditions of migrant workers in China are poor, with small living spaces per person, incomplete residential facilities, and higher safety risks (China Labor Bulletin 2019). With respect to the living conditions of migrant workers at Guiyu town, most of them live in the "old stockade (*laozhai*)" – old traditional houses rented out by local people at a very low price. The environment in the *laozhai* is very poor. The old dilapidated houses inside were initially built around natural rivers and pools when the water was clean; however, nowadays the polluted black water is covered with different kinds of rubbish and this fills the buildings with an unpleasant odor. Tap water is not available in the area of the old stockade. As the ground water is highly polluted, people have to buy and carry water from outside for drinking and cooking purposes. However, most people still use the ground water to wash vegetables, do laundry, and take baths. Figures 5.4 and 5.5 show the outside and inside of a *laozhai* in Guiyu town. Most local people live in either newer *laozhai* or new buildings that have better access to facilities. Many local people have even moved out of the town and into urban cities in order to escape the pollution. Figure 5.6 shows the area of Guiyu where the new buildings, housing local people, are located. Geographically speaking, *laozhai* are closer to environmental pollution and risk than the new buildings that local people live in. Figure 5.7 shows Guiyu town's housing distribution mode. Based on the author's walk-along tours on the ground, this map includes 51 sites of newer *laozhai*, 51 sites of new buildings, and 50 sites of older *laozhai* in 15 villages/communities at Guiyu town,[3] showing

FIGURE 5.4
Appearance of one *laozhai* from the outside (source: taken by the author on March 3, 2018).

FIGURE 5.5
Appearance of one *laozhai* from the inside (source: taken by the author on March 26, 2018).

FIGURE 5.6
New buildings belonging to the local people (source: taken by the author on March 9, 2018).

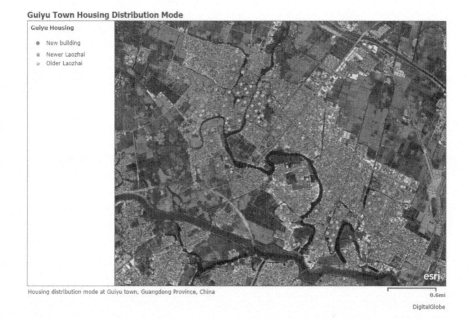

Guiyu Town Housing Distribution Mode

Guiyu Housing

- New building
- Newer Laozhai
- Older Laozhai

Housing distribution mode at Guiyu town, Guangdong Province, China

0.6mi

DigitalGlobe

FIGURE 5.7
Guiyu town's housing distribution mode (created using ArcGIS Online by Esri. ArcGIS® and ArcMap™ are the intellectual property of Esri and are used herein under license. Copyright © Esri. All rights reserved).

that nearly all older *laozhai* are located closer to the polluted river than the other two residence types.

Migrants' living spaces have also been proved to pose higher risks of exposure to damaging chemicals. Even though the living areas of migrant workers are far from their workplace, much higher levels of heavy metals such as lead, copper, and tin are found in the dust in these workers' living spaces than in those of households with no connection to the e-waste industry (Brigden et al. 2005). Workers' home environments are likely being polluted by the chemicals from their workplaces through items like contaminated clothes. An analysis of migrant workers' daily routines (work place–road–home) at Taizhou – a similar e-waste recycling center in China – revealed that the health risks facing migrant workers are 3.8 times greater than those facing local people (Wang et al. 2016). This is a result of a combination of exposure to dangerous chemicals in the workplace and patterns of activity that contaminate their dwelling environments.

Migrant–local relationships and gender are key factors that make female migrants socially and environmentally vulnerable (Wang 2008). Discrimination and exclusion perpetrated by local people toward outsiders are well known. Physical bullying and assault – particularly toward women – for no reason has happened a lot since migrant workers first came to Guiyu in the 1990s. One female migrant worker said, "Both the local old and young people bullied female migrants ... Before, we didn't even dare to walk on the street alone because even the children bullied and beat us." Nowadays, even though physical conflict has become rarer, disrespect and discrimination toward migrants still persists in everyday life. Because migrant workers come from poorer rural areas, they are always labeled as dirty, rude, and annoying with lower *suzhi*.[4] Many local people commented online that migrants are dirty and rude, accusing some of them of stealing, robbing, and disturbing the peace of Guiyu. One local interviewee said, "I personally don't like the female migrants, they are provincial and dirty." The strong local clan system is another reason that leads to discrimination toward and exclusion of migrant people. People in the clan have a strong tendency to help each other within the in-group and unite to fight against outsiders. Migrant people are called "*waishengzai* (people from other provinces)" which has negative connotations in the local dialect. One female teenager said: "They call us '*waishengzai*' ... Our hometown is much better, no one calls us '*waishengzai*' there, so we can easily integrate into the community there."

Exclusion happens not only in daily life but also at an institutional level. As most migrant workers' *huji* remains in their hometown, administratively the local government in Guiyu town does not have responsibility for or jurisdiction of the migrant workers. As a local village committee member said, "There is nothing for us to do with the migrant people administratively ... We don't have much interaction in daily life." In addition, representatives of women's congresses (*funv zhuren*) in town-level governments and village-level committees are supposed to be responsible for conducting female-related work as well as representing and safeguarding the interests of women and children under the leadership of the all-China Women's Federation. However, representatives of the women's congresses in the villages of Guiyu town are mainly responsible for family planning work relating to local people and usually do not have any engagement with migrant women. Politically speaking, migrants do not have the right to vote or stand for election in Guiyu town as they do not belong to these villages. The administrative and political absence of migrant workers means that the interests and rights of migrant workers cannot be represented in decision-making processes and are placed totally outside of collective consideration. For example, *laozhai* as the main living space of migrant workers is almost abandoned and always the last place village committees look to manage. Furthermore, migrants do not have effective channels to protect themselves when their interests are violated, for example if they suffer from injuries at work or receive unfair treatment from the local government.

The exclusion is not only due to the institutional barrier of the *huji zhidu* but also due to the intentional disregard of migrant people by the local governments from a cultural perspective. Migrants have articulated experiences including being subject to arbitrary charges, threatening behavior, inaction, and cover-ups of local people's bullying actions. In addition, migrant workers are faced with significant personal risk as a result of the enforcement of environmental protection laws by the local government. Workers are usually arrested and fined if the owner cannot be found during unannounced inspections of illegal workshops.

Gender and age are closely intersected with migrant status in shaping the experience of social and environmental injustice. People are not necessarily equally influenced by the same environmental pollution (Cutter et al. 2000), with women – especially pregnant women – and children being more environmentally vulnerable in terms of their physical condition. Research has shown that the rate of spontaneous abortion in Guiyu town is much higher than in other places (Wu et al. 2012). Children are arguably the most vulnerable group. Studies have verified that the level of heavy metals in the bodies of most children in Guiyu town exceeds the safe range (Peng 2005; Zeng et al. 2016).

The environmentally vulnerable status of females and children is not only due to their physical vulnerability but also due to the prevailing patriarchal norms and practices. For example, it is women who do most of the housework, which can increase the chance of interaction with polluted substances. In contrast to male migrants, female migrants shoulder the combined burden of working, doing housework, and taking care of children simultaneously. Only in very few migrant families does the husband help with the housework. As one female migrant interviewee said, "The life of the female is definitely harder than that of the male. Females have to do the housework, take care of the children, and work to earn money. It is very tiring. When males are resting in the park and playing cards, females are working outside or doing housework at home. You rarely see females in parks or other public areas."

In addition, the combination of their migrant status, the patriarchal tradition of favoring boys, and the lesser opportunities available to achieve higher education, contributes to the powerlessness of female migrant teenagers. Girls start doing housework and helping with odd e-waste work at home from a very young age. Figures 5.8 and 5.9 show females, including female children, washing clothes in the polluted river. Female migrant children are also excluded from the education system because of their migrant status and the patriarchal tradition. With the exception of some children who stay and study in their hometowns, migrant children usually attend private schools that are cheap and which have relatively few educational resources. In addition, as a result of cultural norms favoring boys, girls usually do not benefit from as many resources as boys in the family do, and most leave school and start working early.

The Working Environment

The working conditions of migrant workers in Guiyu town are rough and harsh. Figure 5.10 shows the basic working environment in a family workshop. Migrant workers work without any effective protection against waste air and water. The working space is filled with the strong smells of burning plastics and other e-waste. One of the most typical jobs – undertaken

FIGURE 5.8
Females washing clothes along the polluted river (source: taken by the author on March 4, 2018).

mostly by women – in the industry is melting circuit boards to get the metal parts. Workers sit directly in front of the stove, which is heated to over 400 degrees Celsius, with only a small fan above their heads. Both the air and water waste in these workshops have been proven to contain toxic heavy metals and organic compounds that could cause irreversible harm to human health (Brigden et al. 2005; Qiu et al. 2005).

Power in the workplace particularly manifests in the relationship between the workshop owners and the migrant workers. Workshop owners use a range of different techniques to manipulate and exploit laborers, ranging from coercive strategies of threats, punishment, and even assault to soft power strategies of treating migrants well, building dependency, and encouraging a long-term cooperative relationship. On the one hand, tricks, regulations, threats, and punishments are often used to help cultivate an atmosphere of fear and discipline in the workplace. Women are particularly vulnerable to such coercive power. Security cameras are installed at every corner of the workshop and the local owner/overseer regularly walks around to inspect the workers. These measures help generate conditions described by Foucault's (1991) panopticon observation, whereby workers under an all-perceiving eye discipline themselves. As an interviewee illustrated to me: "When he (the boss) stares at me, I feel afraid … I can see his shadow through the light and know that he stands still right behind me, sometimes even for

FIGURE 5.9
Female children washing clothes along the polluted river (source: taken by the author on March 4, 2018).

half an hour. I didn't even dare to look up." In addition, nearly all migrant workers are employed through informal systems and do not have any signed contracts with their employers. It is normal practice that owners do not make the required social insurance payments in relation to migrant workers, including medical insurance and work-related injury insurance. The prevalence of informal employment is related to the instability and fluidity of the jobs, due in turn to the inconsistent supply of e-waste resources.

On the other hand, the "benefit strategy" is often mentioned by migrant workers. Workshop owners who use this strategy consider it important to maintain the "consent" of the migrant workers toward the work. Through cultivating dependency, this strategy shapes the perceptions of workers and generates a sense of loyalty toward their boss. According to the author's fieldwork experience, such a strategy is more likely to be accepted by females. As one female migrant interviewee explained, "The boss treated me very well at that time ... We came here because we couldn't survive in our hometown. When I first came here, we were so poor and couldn't even afford to eat meat. My boss accepted me, gave me a job, and taught me how to work. He also often gave me leftovers, cookies, and candies ... He told me not to tell other people my salary."

Conceptions about female migrant workers and the working environment also help to justify exploitation. The harsh working environment and the

FIGURE 5.10
Inside of a family workshop (source: taken by the author on April 7, 2018).

difficulty of doing e-waste jobs have been widely normalized. When asked about the dirty working environment and its influence on the workers, one workshop owner responded, "What job in the Industry Park is not dirty? There is no clean work in the Industrial Park and Guiyu town ... All the jobs use life in exchange for money" (Chen et al. 2014). The capacity to endure the terrible environment and hardship is further whitewashed as a kind of hard-working "spirit" that is necessary if ones hopes to earn money and achieve success. Migrant workers themselves even accept such conceptions, regarding the endurance of hardship as an essential stage of eventual success. For example, one female migrant worker said, "We have already got accustomed to the bad environment and the hardship. Sometimes we also feel proud of enduring this kind of hardship." On the other hand, migrant workers are often pictured as willing to choose e-waste jobs and caring only about earning money, a depiction used to justify social and environmental exploitation of the workers. This leads to ignorance of migrant workers' right to work in a clean and healthy environment and further the lack of recognition and devaluation of migrant workers. Another migrant worker told the author, "They don't care about how old you are or other things, they only need labor to work for them ... If you are a migrant worker and want to find a job, no one respects you."

In addition, migrant status and class are closely related and even bonded together, shaping the perception of migrant workers as being willing to choose unhealthy jobs and sacrifice their health to earn money. As we have seen in the above section, migrant workers are stripped of the right to be protected from environmental pollution just because of their migrant status – they are fluid, willing to do such jobs, and do not belong to Guiyu town culturally, administratively, or politically. "Migrant" and "workers" are bonded so closely that the two concepts form an infinite loop – it is because migrant people are regarded as having low *suzhi* that they are humble enough to accept dirty e-waste jobs, and it is because they are e-waste workers that they are less respected, recognized, and valued as migrants.

Power based on gender and age also manifests in the workplace, intersecting with power based on class and migrant status. There is thus an unbalanced gender and age division of labor in the working environment. Gender combined with age has determined certain working patterns in the labor market. While male migrants and young female migrants usually have more choices and opportunities to enter other industries with better conditions, older female migrants and children usually cannot select jobs at will. For example, a lot of male migrants choose jobs that involve physical labor, such as transporting and building, instead of working in e-waste workshops. Large-scale industries, such as textile mills, prefer to employ younger females. This is based on the perception that older people learn things more slowly than young people, and that older people can endure more hardship and dirty work so they are suitable for e-waste dismantling work. An older female migrant worker said, "There are very few older people entering large-scale industries. All are young people. Because [they think] older people are too stupid. Even though you are willing to learn, they still don't employ you." In addition, even for the same work, female workers' salaries are lower than male workers'. Therefore, to earn more money, female migrants have to do jobs that are dirtier and unhealthier such as burning circuit boards, which attracts a relatively high salary compared to other easier work.

Hiring child labor is also quite normal in Guiyu town. Even during the school term, many female migrant children work part-time jobs at individual workshops. Some work in family workshops to earn money for their family during weekends and holidays, while others do odd jobs together with their parents at home. In addition, the conception that girls do not need to study and should go to work to earn money occupies a lot of migrants' minds. Some migrant girls themselves favor the idea of quitting school and earning money as early as possible. A female teenager said, "My performance is not good at school, I don't want to read anymore. I want to earn money as soon as possible, then I will be free."

Both in the period when serious environmental pollution was commonplace and in the contemporary period of increased environmental governance, migrant workers' environmental suffering and interests have been ignored by both the central and the local governments. For example, the working environment of family workshops (both inside and outside of the Industrial Park) has not been improved at all and conversely has become worse than before. A female migrant worker said, "The Industrial Park looks good from the outside, but workers suffer more than before because of the substandard environmental protection facilities. In previous family workshops, the air flowed more easily." Migrant workers' suffering has never been constructed as a problem that needs to be attended to and dealt with. In addition, gender is the social difference that gains the least attention from the public. If we say migrant workers are "beneath the shadow of a problem," female migrant workers are not even recognized in the existing stories of migrant workers.

Conclusion

It has been argued that the process of migration alleviates the pressure on migrant women from their parents-in-law in terms of being far away from the "father power" (e.g., Ma 2003; Tan 2004). A lot of migrant women do become more empowered and independent in the family as they start earning money. However, the experience of migration – intersected with remained patriarchal practices and perceptions – can result in worse experiences for females because they have less chance to achieve a good education; they are more exposed and vulnerable to environmental pollution; and they will likely face a more difficult life in the future. Additionally, female migrant workers' independence and empowerment is achieved at the cost of the devaluation and misrecognition of their labor both as migrants and as women.

All in all, we can see that different dimensions of power based on class, migrant status, gender, and age are intersected as a network to produce the social and environmental injustice experienced by female migrant workers. This results in the maldistribution of environmental pollution, the exclusion of female migrant workers from governance and decision-making processes, the misrecognition and devaluation of female migrant workers, and the maintenance of barriers restricting their capacity development. The qualitative GIS approach used in this chapter allows us to have a more contextual understanding about the intersected power relations and the lived experiences of female migrant workers. It further shows that GIS applications have the potential to bridge the gap between qualitative and quantitative analyses that often results in the ignorance of the environmental injustice faced by females.

Notes

1. *Huji zhidu* is required by law in mainland China and determines where citizens can live. It is an important tool to help control migration into urban areas and maintain social stability.

2. It is important to note that it is nearly impossible to ascertain the accurate demo-graphic composition of these migrant workers and workshop owners, even when using official documents, other publications/resources, and the author's own fieldwork. The local government's data on migrant workers are based on regis-trations of migrant people at the local police station. This is not accurate because only a few migrant workers have registered. Another reason is that this popula-tion and the work they do are highly fluid, meaning it is difficult to get a fixed sense of this population in this area. The official number of migrant workers was 50,000 in 2011. However, Frey (2012) indicates in his research that by 2012 there were around 150,000 migrant workers. According to estimations from different reports and studies, there were even more migrants than local people – about 200,000 nowadays – during the heyday of the e-waste industry.
3. There are 27 villages/communities in Guiyu town. The author has only visited the villages that are most closely involved in the e-waste recycling industry.
4. *Suzhi* (personal quality) refers to the concrete manifestation of a person's thoughts and behaviors in social life. It involves a set of attributes including education, morality, physiology, thinking and acting ability, occupational skills, etc. Developing comprehensive quality (*zonghe suzhi*) is an important part of modernization in post-Mao era.

References

Brigden, K., I. Labunska, D. Santillo, and M. Allsopp. 2005. Recycling of electronic wastes in China and India: Workplace and environmental contamination. *Greenpeace International* 55. . https://www.greenpeace.org/international/publ ication/7051/recycling-of-electronic-waste-in-india-and-china-summary/

Census office of the State Council, and Population and Employment Statistics Division of the National Bureau of Statistics. 2010. *Tabulation on the 2010 popula-tion census of the people's republic of China*. China Statistic Press and Beijing Info Press. http://www.stats.gov.cn/tjsj/pcsj/rkpc/6rp/indexch.htm.

Chen, J. 2011. Internal migration and health: Re-examining the healthy migrant phe-nomenon in China. *Social Science & Medicine* 72 (8): 1294–1301.

Chen, J., S. Chen, and P. F. Landry. 2013. Migration, environmental hazards, and health outcomes in China. *Social Science & Medicine* 80: 85–95.

Chen, J., S. Li, and X. X. Kong. 2014. The capital of e-waste is struggling with trans-formation. *Economic Information Daily*, July 24. http://jjckb.xinhuanet.com/201 4-07/24/content_513985.htm

China Labour Bulletin. 2019. *Migrant workers and their children*. https://clb.org.hk/c ontent/migrant-workers-and-their-children

Cutter, S. L., J. T. Mitchell, and M. S. Scott. 2000. Revealing the vulnerability of people and places: A case study of Georgetown County, South Carolina. *Annals of the Association of American Geographers* 90 (4): 713–737.

Foucault, M. 1991. *Discipline and punish: The birth of a prison*. London: Penguin.

Hanser, A. 2008. *Service encounters: Class, gender, and the market for social distinction in urban China*. California: Stanford University Press.

Hope Alkon, A. 2011. Reflexivity and environmental justice scholarship: A role for feminist methodologies. *Organization & Environment* 24 (2): 130–149.

Jacka, T. 2014. *Rural women in urban China: Gender, migration, and social change.* New York: Routledge.

Krauss, C. 1993. Women and toxic waste protests: Race, class and gender as resources of resistance. *Qualitative Sociology* 16 (3): 247–262.

Li, Q. 2004. *Migrant workers and social stratification in China.* Beijing: Literature of Social Science Press.

Li, Y. S., and Y. X. Liu. 2006. Social policy and social protection towards migrant workers. *Social Science Research* 6:100–105.

Ma, C. B. 2010. Who bears the environmental burden in China – An analysis of the distribution of industrial pollution sources?. *Ecological Economics* 69 (9): 1869–1876.

Ma, C. H. 2003. *Marketization and gender relations of rural families in China.* PhD diss., Graduate School of Chinese Academy of Social Sciences.

Nie, W. 2013. Social-economic status and environmental risk distribution: Based on the quantitative study of Xiamen waste disposal. *Journal of China University of Geosciences (Social Sciences Edition)* 13 (4): 45–52.

Ong, A. 1991. The gender and labor politics of postmodernity. *Annual Review of Anthropology* 20 (1): 279–309.

Peng, L., X. Huo, X. J. Xu, Y. Zheng, and B. Qiu. 2005. Effects of electronic waste recycling disposing contamination on children's blood lead level. *Journal of Shantou University Medical College* 18: 48–50.

Pun, N. 2005. Made in China: Women factory workers in a global workplace. Durham: Duke University Press.

Qiu, B., L. Peng, X. J. Xu, et al. 2005. Investigation of the health of electronic waste treatment workers. *Journal of Environment Health* 22 (6): 419–421.

Schlosberg, D. 2009. *Defining environmental justice: Theories, movements, and nature.* New York: Oxford University Press.

Schoolman, E. D., and C. B. Ma. 2012. Migration, class and environmental inequality: Exposure to pollution in China's Jiangsu province. *Ecological Economics* 75: 140–151.

Frey, R. S. 2012. The e-waste stream in the world-system. *Journal of Globalization Studies* 3 (1): 79–94.

Sun, X. L., and R. H. Shi. 2016. Residential segregation and community exposure risk: Environmental inequality in megacity. *Journal of CAG* 6: 86–93.

Tan, S. 2004. Family strategy or individual autonomy? – Gender analysis of the decision model of rural labor migration. *Zhejiang Academic Journal* 5: 210–214.

Teixeira, S. 2018. Qualitative geographic information systems (GIS): An untapped research approach for social work. *Qualitative Social Work* 17(1): 9–23.

Tschakert, P. 2009. Digging deep for justice: A radical reimagination of the artisanal gold mining sector in Ghana. *Antipode* 41(4): 706–740.

Walker, G. 2012. *Environmental justice: Concepts, evidence and politics.* Oxon; New York: Routledge.

Wang, S. M. 2008. The bottom line of the right to live, environmental right and social exclusion – a sociological perspective on the experience of environmental justice. In Xu, X. M. (ed.). 2008*China Environment and Resources Law Review (Vol. 2007)* (pp. 292–305). Beijing: People's Publishing House.

Wang, Y. L., J. X. Hu, W. Lin, et al. 2016. Health risk assessment of migrant workers' exposure to polychlorinated biphenyls in air and dust in an e-waste recycling area in China: Indication for a new wealth gap in environmental rights. *Environment International* 87: 33–41.

Wu, K. S., X. J. Xu, L. Peng, et al. 2012. Association between maternal exposure to perfluorooctanoic acid (PFOA) from electronic waste recycling and neonatal health outcomes. *Environment International* 48: 1–8.

Zeng, X., X. J. Xu, H. M. Boezen, and X. Huo. 2016. Children with health impairments by heavy metals in an e-waste recycling area. *Chemosphere* 148: 408–415.

6

Spatial Concentration of Social Vulnerability and Gender Inequalities in Mexico

Roberto Ariel Abeldaño Zuñiga and Javiera Fanta Garrido

CONTENTS

Introduction

Since the beginning of the 21st century, social vulnerability has gained increasing attention among researchers and policymakers from Latin America and the Caribbean, as social well-being in the region tends to be fragile and unstable. Although there are multiple definitions of social vulnerability, the concept is usually associated with perceptions of uncertainty and helplessness regarding factors that influence the living conditions of the population (Pizarro 2001; Stampini et al. 2016). These factors are generally related to access to work, housing, social security, and the possibilities of consumption and credit. While the restrictions for accessing these factors are not new for the poorer sectors of the population, the lower middle and middle classes of the region represent increasingly wide margins of probability of fluctuating quality of life, whether due to upward or downward social mobility (Busso 2001; Sánchez and Egea 2011; Rubio Herrera and Flores Palacios 2018). The effects of the current pattern of development, structural adjustment policies, which have deepened social inequalities, and the growing lack of social protection are some of the factors that explain this growing risk.

This work is based on a realistic paradigm of the notion of vulnerability (Ruiz Rivera 2012). This paradigm emphasizes the physical threats, objective risks, and social conditions that underlie the processes of social inequality and the assessment that each group contributes to the risks. Based on the construction of a synthetic index, the objective of this research is to examine the degree of social vulnerability of the Mexican households and identify where the hot and cold spots of social vulnerability are located (Anselin 2019). To achieve this objective, a geospatial analysis among Mexican municipalities was applied and the degree of social vulnerability was compared between female- and male-headed households.

In this work, vulnerability is understood as the inability of an individual, household, or social group to face, neutralize, or obtain benefits from these risks (Pizarro 2001). This definition involves two main components: (1) the insecurity experienced by communities, homes, and individuals in their living conditions due to the catastrophic or chronic effects of an event or situation and (2) the availability and management of resources and strategies to deal with the consequences generated by these events or situations.

Within the specialized literature, the notion of vulnerability is commonly delimited to specific areas (e.g., vulnerability to climate change, mortality due to certain causes, and food insecurity, among other aspects). In each of these fields, vulnerability involves a multidimensional concept, due to the different levels of well-being of an individual, home, or community that may be affected by objective risks or threats. In methodological terms, the measurement of this concept usually focuses on two specific aspects (Ruiz Rivera 2012): (1) the intensity of the danger to which a person or group is exposed and (2) the objective material conditions that originate, favor or intensify these dangers. This work focuses on the second aspect mentioned, considering households as the unit of analysis. Various authors (Huffman et al. 2019; International Labour Office 2003; Sinding 2009; Merrick 2002; Narayan 2002) have demonstrated the usefulness of the household approach. Households operate as a basic organizational unit for survival and daily reproduction, and the modality of coexistence among their members can determine the access and variety of resources. It has been observed, for example, that households with more than one adult deploy more specialized strategies to face uncertainties related to the economy, or to mitigate the effects of poverty and destitution, compared to households with just one adult (Huffman et al. 2019).

Measuring Social Vulnerability in the Mexican Context

Considerable research has been published with respect to social vulnerability in Mexico. The most common topics examined are vulnerability to climate change (Ibarrarán et al. 2008; Esperón-Rodríguez et al. 2016; Soares and Gutiérrez 2011; Soares and Murillo-Licea 2013), environmental hazards (Zúñiga et al. 2014; Pérez-Maqueo et al. 2018; Neri and Magaña 2016), food insecurity (Vilar-Compte et al. 2015; Martínez-Rodríguez et al. 2015), and adolescent pregnancy and motherhood (Stern 2012; Villalobos-Hernández et al. 2015). The gender approach has been increasingly included in these studies, as there is widespread agreement that gender inequalities are a key component of social vulnerability. Indeed, gender inequalities determine the intensity of the risks, the way they are perceived, and the response capacity and coping strategies related to each problem or event (Soares and Murillo-Licea 2013).

This work proposes the construction of a synthetic index applicable to the Mexican context, which articulates variables for the measurement of social vulnerability in households. All of these variables are included in the database of the Inter-Census Survey that the National Institute of Statistics of Mexico has completed with a sample representative of the national domain in 2015 (Instituto Nacional de Estadística y Geografía 2015b, 2015a).

Although not all aspects of vulnerability are observable through the available sources, a substantial detailed information of individuals and households are present in the census data sources, which allows the identification of some groups in situations of social risk and disadvantage. Taking into account this methodological and conceptual precaution, the most significant dimensions of the social vulnerability of households were reconstructed according to the methodology suggested by Melina Con (2009), considering the indicators mentioned below.

The material assets dimension includes four basic indicators: (1) Household crowding; (2) Quality of housing materials (considering material quality of floors, walls, and ceilings); (3) Dependency burden of household income earners; and (4) Access to basic services within the home (access to drinking water, electricity, drainage, and cooking fuel).

The non-material assets dimension includes two basic indicators: (1) Health insurance of the head of household and (2) Number of years of formal education of the head of household.

To articulate all dimensions in an integral index, a weighting structure was created by applying weights to variables, based on criteria defined in the literature (Con 2009; Abeldaño et al. 2013). Table 6.1 shows the weighting structure for all the variables included.

By integrating the six dimensions in a final indicator, a social vulnerability index (SVI) was obtained for each household, whose value can range between 0 and 1, where 0 represents the absence of social vulnerability and 1 implies extreme social vulnerability. The categories within each dimension are mutually exclusive. In the data processing, households without information on any of the variables involved in the processing and households residing in collective institutions were excluded.

Following Roche's methodological approach (Roche 2008), a total of 5,770,539 households that met the above criteria were analyzed. To assess the validity of the index, an exploratory factor analysis was performed by entering the variables that make up each of the indicators constituting the index. The KMO sample adequacy test was highly satisfactory, with a coefficient of 0.937, while Bartlett's sphericity test was significant ($p<0.01$). These

TABLE 6.1

Variables Included in the Construction of the Social Vulnerability Index, with Their Respective Weights

Dimensions	Variables	Weight
Overcrowding	Non-overcrowded home, up to 2 persons per room	0.00
	Overcrowded home, more than 2 and up to 3 persons per room	0.10
	Overcrowded home, more than 3 persons per room	0.15
Material quality of housing	The house has no deficiency in its construction materials	0.00
	The house has at least one deficiency in its construction materials	0.10
	The house has two deficiencies in its construction materials	0.15
	The house has three deficiencies in its construction materials	0.20
Access to basic services in housing	The house does not lack any service	0.00
	The house lacks one service	0.05
	The house lacks two services	0.10
	The house lacks three services	0.15
	The house lacks four services	0.20
Dependency burden of income earners	Household with 5 or more members and the head of household is retired; in addition the household does not receive money from other people inside or outside the country, or from government programs	0.20
	Household with 2 or more members and the head of household is neither working nor retired; in addition the household does not receive money from other people inside or outside the country, or from government programs	0.20
Health Insurance of head of household	The head of household has health insurance	0.00
	The head of the household has no health insurance	0.10
Educational level of head of the household	The head of household has 12 or more years of schooling	0.00
	The head of household has between 7 and 12 years of schooling	0.10
	The head of household has up to 7 years of schooling	0.15

coefficients were adequate to conduct the exploratory factor analysis. The technique used was that of main components, with a forced extraction of three factors, which together accounted for 78.3% of the total variance of the data (figure 6.1). Varimax was the chosen rotation.

The sedimentation graph shows the eigenvalues in the y-axis and positions the components in the x-axis. The graphic layout – the changes in the slope – helps to see how much explanatory capacity each component provides. This graph shows the optimal number of factors and graphically represents the size of the eigenvalues ordered from highest to lowest. A self-value indicates the amount of variance explained by a main component. If a self-value approaches zero, it means that the factor corresponding to that self-value is unable to explain a relevant amount of the total variance. Therefore, a factor

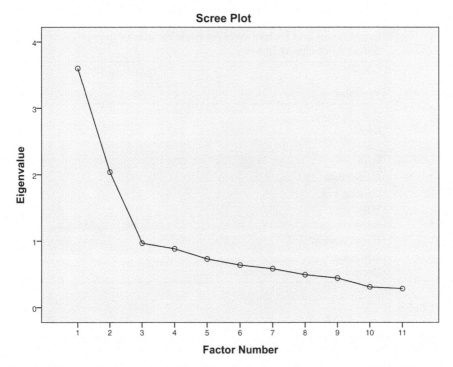

FIGURE 6.1
Components of the social vulnerability index.

that corresponds to an auto-value close to zero is considered a residual and meaningless factor in the analysis. In the graph the slope loses inclination from the fourth self-assessment, so it is considered that only the first three factors should be extracted and discarded from the fourth onward.

Table 6.2 shows the components of the SVI rotated in space. From this analysis, three dimensions of grouping of the variables entered into the analysis that respond to three theoretical dimensions, namely, housing structure, basic housing services, and social assets of household heads. Here we should note that the variable "drainage" presented high factor loads in two components, but by theoretical consistency we then decided to include it within the "housing services" dimension.

Social Vulnerability and Gender Inequalities in Mexico

This section analyzes the social vulnerability presented by Mexican households, according to the following types of households: (1) All households in the country, (2) Households consisting of adults only (with no presence of

TABLE 6.2

Components and Factor Loads for Each Variable of the
Social Vulnerability Index

	Component		
Variables	1	2	3
Sleeping rooms	.857		
Roofs	.813		
Floors	.758		
Walls	.736		
Electric light		.642	
Cooking fuel		.535	
Water		.532	
Sewer system		−.499	.585
Health services			.545
Education			.450
Dependency burden of income earners			.410

TABLE 6.3

Mean Values of the Social Vulnerability Index of Mexican Households

		C.I. 95%		
Household Typology	Mean	Lower	Upper	Std. Error
Households composed of adults only	0.2190	0.2171	0.2210	0.0010
Households with presence of children under 18	0.2684	0.2661	0.2707	0.0012
Male-headed households	0.2492	0.2468	0.2516	0.0012
Female-headed households	0.2527	0.2507	0.2546	0.0010
Male-headed households without children	0.2167	0.2147	0.2188	0.0010
Male-headed households with children	0.2656	0.2631	0.2680	0.0012
Female-headed households without children	0.2237	0.2219	0.2255	0.0009
Female-headed households with children	0.2767	0.2746	0.2787	0.0010
All households	0.2503	0.2481	0.2526	0.0011

children under 18 years old), (3) Households with children under 18 years old, (4) Male-headed households, (5) Female-headed households, (6) Male-headed households without presence of minor children, (7) Male-headed households with presence of minor children, (8) Female-headed households without presence of minor children, and (9) Female-headed households with presence of children under 18 years of age.

Table 6.3 shows that households with the greatest social vulnerability are those made up of a female head of household with minor children. This finding could be due to the dependence burden of minors on women receiving an income. Subsequently, households with the presence of minors are those with higher social vulnerability. Female-headed households showed greater vulnerability than male-headed households. This observation leads to a clear pattern: female-headed households and female-headed households with the presence of minors are those with the greatest social disadvantages.

To be able to visualize the spatial patterns, the household level data were aggregated to the total of 2,418 municipalities of Mexico. The average social vulnerability index of households by municipality was calculated to apply a spatial self-correlation analysis using Moran's I statistic.

In the analysis of spatial self-correlation of all households in Mexico, 1,002 statistically significant municipalities were found with a Moran's I of 0.43, which means that the pattern of spatial distribution was not random. Of those 1,002 municipalities, a concentration zone of 342 classified as hot spots (high vulnerability), since they were municipalities with high social vulnerability index values, significantly surrounded by other municipalities with the same characteristics; this mainly affects the states of Oaxaca, Chiapas, and Guerrero. There were also cataloged 36 municipalities in the north of the country that had high index values, surrounded by other municipalities with low values. The conglomerates of municipalities with low levels of social vulnerability cold spots (low vulnerability) were located in the north of the country (figures 6.2a and b).

In the analysis of households composed by adults, 977 statistically significant municipalities were identified with a positive spatial autocorrelation (Moran's I = 0.44). The number of municipalities identified as hot spots was 367, distributed among the states of Oaxaca, Chiapas, Guerrero, Michoacán, Puebla, Tlaxcala, and Veracruz. In northern Mexico, municipalities that have

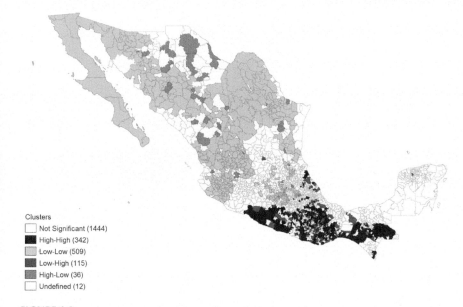

Clusters
☐ Not Significant (1444)
■ High-High (342)
▨ Low-Low (509)
■ Low-High (115)
■ High-Low (36)
☐ Undefined (12)

FIGURE 6.2
a: Spatial clusters in all households, by municipalities.

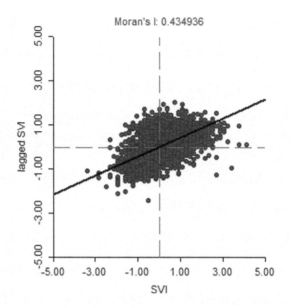

FIGURE 6.2
b: Scatter plot of Moran coefficient I.

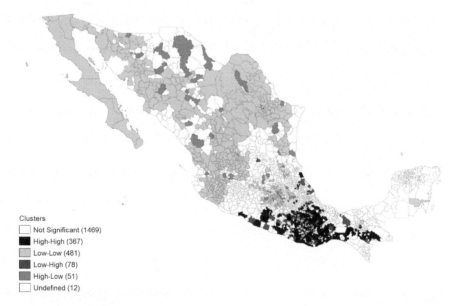

FIGURE 6.3
a: Spatial clusters in all households composed by adults, by municipalities.

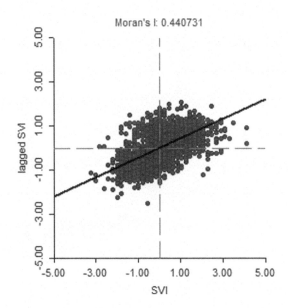

FIGURE 6.3
b: Scatter plot of Moran coefficient I.

high values in the index were also identified, significantly surrounded by other municipalities with low values (figures 6.3aand b).

In the stratum formed by households with presence of minors, 1,016 statistically significant municipalities were identified, with a positive spatial autocorrelation (Moran's I = 0.44). 274 municipalities were identified as hot spots (highly vulnerable) in the south of Mexico (Oaxaca, Chiapas and Guerrero states) (figures 6.4a and b).

In the stratum of male-headed households, 928 statistically significant municipalities were identified, with a Moran's I correlation of 0.42. 358 municipalities were identified as hot spots in the states of the south region (Oaxaca, Chiapas, Guerrero) and in the state of Veracruz. In northern Mexico, municipalities that have high values in the index were also identified, surrounded significantly by other municipalities with low values. (figures 6.5a and b).

In the analysis of the female-headed households, 928 statistically significant municipalities were identified, with a Moran I correlation of 0.42. 322 municipalities were identified as hot spots. Although the number of municipalities with highly vulnerable households is less, they are found in more states: Oaxaca, Chiapas, Guerrero, Veracruz, and the Yucatan Peninsula; this finding suggests that these states share common policies or historical conditions leading to highly vulnerable households affecting, mainly the female-headed. In northern Mexico, municipalities that have high values in the index were also identified, surrounded significantly by other municipalities with low values (figures 6.6a and b).

In the male-headed households without the presence of minors, 954 statistically significant municipalities were identified, with a Moran I

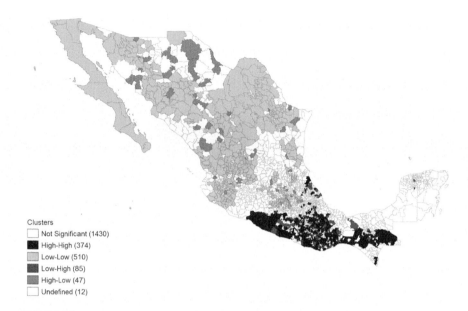

Clusters
- ☐ Not Significant (1430)
- ■ High-High (374)
- ☐ Low-Low (510)
- ☐ Low-High (85)
- ☐ High-Low (47)
- ☐ Undefined (12)

FIGURE 6.4
a: Spatial clusters in households with presence of minors, by municipalities.

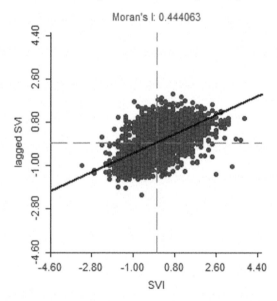

FIGURE 6.4
b: Scatter plot of Moran coefficient I.

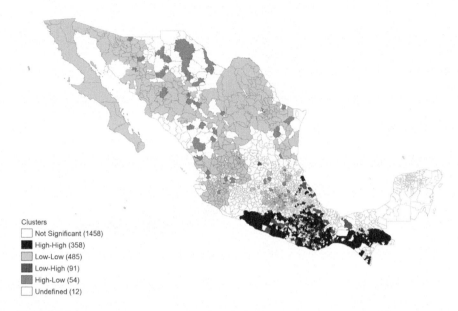

FIGURE 6.5
a: Spatial clusters in male-headed households, by municipalities.

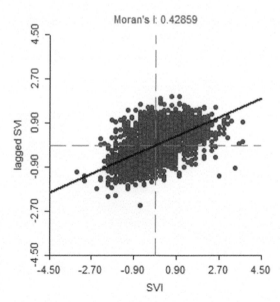

FIGURE 6.5
b: Scatter plot of Moran coefficient I.

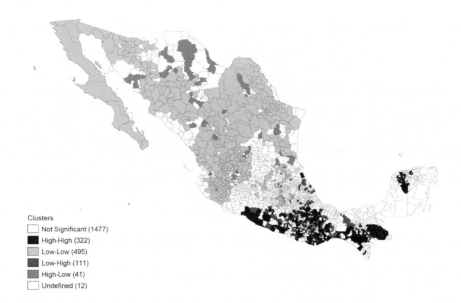

Clusters
☐ Not Significant (1477)
■ High-High (322)
▦ Low-Low (495)
■ Low-High (111)
■ High-Low (41)
☐ Undefined (12)

FIGURE 6.6
a: Spatial clusters in female-headed households, by municipalities.

correlation of 0.42. 357 municipalities were identified as hot spots in the southern states (Oaxaca, Chiapas, Guerrero) and the state of Veracruz. In northern Mexico, municipalities that have high index values were also identified, surrounded significantly by other municipalities with low values (figures 6.7a and b).

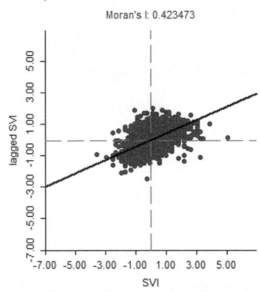

FIGURE 6.6
b: Scatter plot of Moran coefficient I.

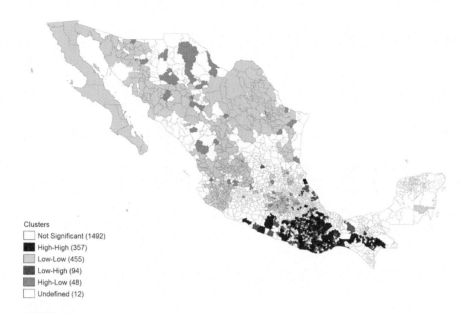

Clusters
◻ Not Significant (1492)
■ High-High (357)
▨ Low-Low (455)
■ Low-High (94)
▨ High-Low (48)
◻ Undefined (12)

FIGURE 6.7
a: Spatial clusters in male-headed households without presence of minors, by municipalities.

In the analysis of the stratum formed by male-headed households with the presence of minors in the home, 928 statistically significant municipalities were identified, with a Moran I correlation of 0.42. 370 municipalities were identified as hot spots in the southern states (Oaxaca, Chiapas, Guerrero) and in the state of Veracruz. In northern Mexico, municipalities that have high values in the index were also identified, surrounded significantly by other municipalities with low values (figures 6.8a and b).

In the stratum formed by female-headed households without the presence of minors, 987 statistically significant municipalities were identified, with a Moran I correlation of 0.44. 336 municipalities were identified as hot spots in the southern states, as well as in Puebla, Tlaxcala, and the Yucatan Peninsula. In northern Mexico, municipalities that have high index values were also identified, surrounded significantly by other municipalities with low values (figures 6.9a and b).

In the last stratum, formed by female-headed households with presence of minors, 964 statistically significant municipalities were identified, with a Moran I correlation of 0.41. 347 municipalities were identified as hot spots in the Southern states, as well as in Puebla, Tlaxcala, and the Yucatan Peninsula. In northern Mexico, municipalities that have high index values were also identified, surrounded significantly by other municipalities with low values (figures 6.10a and b).

Table 6.4 shows the hotspots of social vulnerability among Mexican households and selected socio-demographic features, in the ten municipalities that

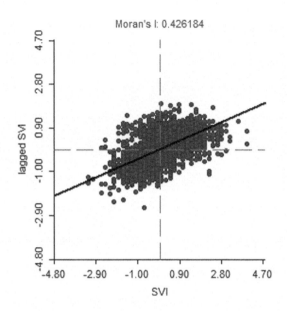

FIGURE 6.7
b: Scatter plot of Moran coefficient I.

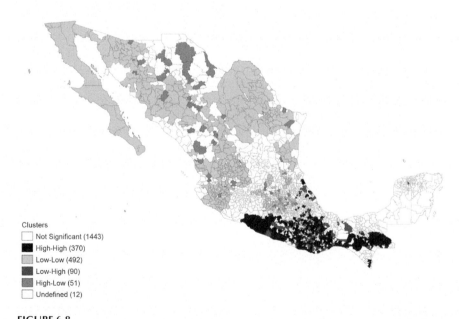

FIGURE 6.8
a: Spatial clusters in male-headed households with presence of minors, by municipalities.

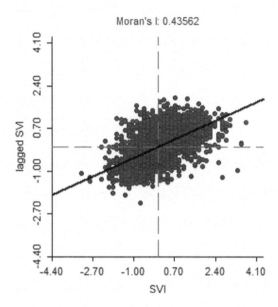

FIGURE 6.8
b: Scatter plot of Moran coefficient I.

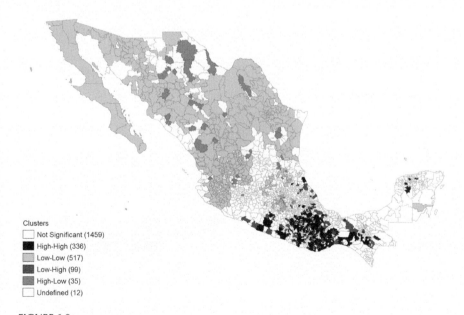

FIGURE 6.9
a: Spatial clusters in female-headed households without presence of minors, by municipalities.

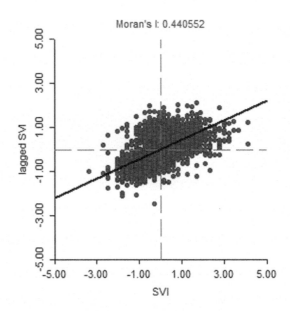

FIGURE 6.9
b: Scatter plot of Moran coefficient I.

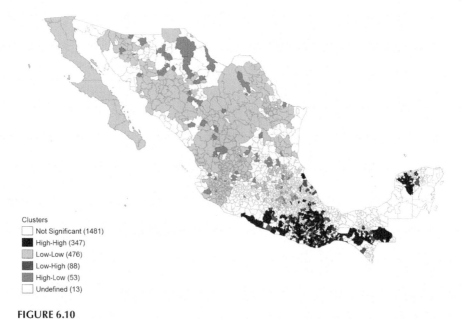

FIGURE 6.10
a: Spatial clusters in female-headed households with presence of minors, by municipalities.

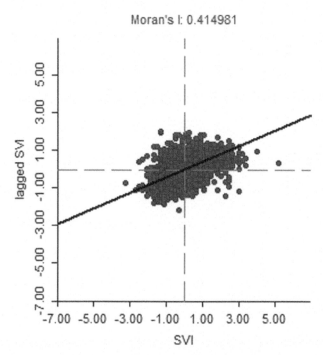

FIGURE 6.10
b: Scatter plot of Moran coefficient I.

showed the highest degree of social vulnerability, with SVI values ranging from 0.40 to 0.46. Overall, those ten municipalities are in the states of Oaxaca, Chiapas, and Guerrero (figure 6.11). As expressed in the geospatial analysis, the hotspots are mainly distributed in the Southern region of Mexico. Noteworthy is the fact that these jurisdictions show percentages of indigenous population above 95% and significant percentages of illiterate population and non-salaried workers. These results are consistent with the fact that at least three-quarters of the households located in these municipalities receive subsidies from the government.

To highlight the results of table 6.4, figure 6.12 shows 570 Oaxaca municipalities according to spatial clusters of social vulnerability and percentage of indigenous population (per municipality), where the consistency between the high percentage of indigenous population and clusters of high social vulnerability is evidenced.

Conclusions

Given its transversal nature, gender gaps prevail in a wide range of situations, events and living conditions. This work attempted to provide an

TABLE 6.4

Social Vulnerability Index (SVI) and Selected Socio-Economic Variables in the Most Vulnerable Municipalities of Mexico

Municipality	State	SVI	Population	% Illiteracy	% Indigenous population	% Non-salaried workers	% Households receiving government social programs
San José Tenango	Oaxaca	0.46	18,316	38.80	98.59	66.74	80.60
Cochoapa el Grande	Guerrero	0.43	18,458	56.13	99.05	75.61	87.40
Coicoyán de las Flores	Oaxaca	0.42	9,936	48.31	98.38	86.40	81.50
Santiago Amoltepec	Oaxaca	0.42	12,683	26.47	96.17	67.93	90.30
San Juan Cancuc	Chiapas	0.41	34,829	34.29	99.04	93.65	84.10
Huautepec	Oaxaca	0.41	6,299	42.45	97.89	42.52	83.90
San Mateo del Mar	Oaxaca	0.41	14,835	20.57	98.87	59.14	75.70
Santiago Nuyoó	Oaxaca	0.41	1,820	19.07	98.74	86.09	82.40
San José Independencia	Oaxaca	0.40	3,867	27.79	99.20	67.81	80.40
Santa María Peñoles	Oaxaca	0.40	8,593	18.89	96.79	69.88	82.90

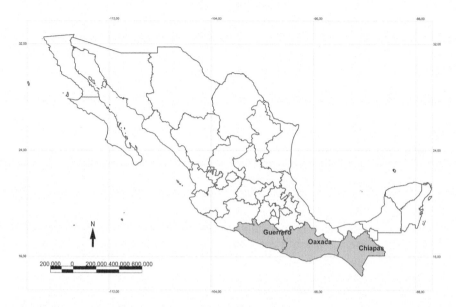

FIGURE 6.11
States of Guerrero, Oaxaca, and Chiapas. Mexico.

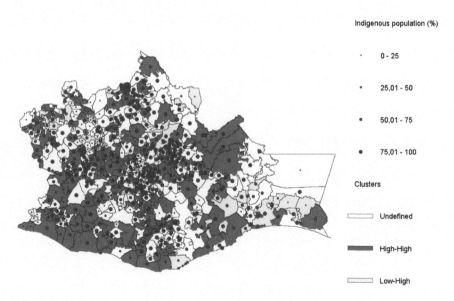

FIGURE 6.12
Clusters of social vulnerability and percentage of indigenous population in the state of Oaxaca, Mexico.

approximation to gender inequalities at the household level, based on the analysis of the degree of social vulnerability to which Mexican households are exposed. The results presented herein lead us to discuss at least two key points.

The first point is the interpretation that is given to the most socially vulnerable group identified by this study, that is, the households headed by women with children. Does this put single mothers and their children as a non-desirable unit for daily reproduction? As a counterpart, are male-headed households and households without children the most successful domestic units, in terms of survival strategies and well-being? It must be noted that due to the characteristics of the available data sources, the concept of social vulnerability developed in this chapter refers to the housing conditions and the main socio-demographic features of the household members. Thus, psychosocial factors such as the existence of community networks and institutional links that could contribute to mitigate or neutralize the impact of risk events and adverse living conditions are not included in the construct. It is possible that female-headed households perform better in this latter dimension, which may delve into decreased gender disparities of social vulnerability. Hence, if progress toward achieving gender equality in the labor market, the recognition and payment of caregiving tasks, or the possibility of diversifying productive resources was made, would not the female-headed households with children be the desirable alternative of daily organization and reproduction?

References

Abeldaño, Roberto Ariel, Juan Carlos Estario, and Alicia Ruth Fernández. 2013. "Analysis and Spatial Distribution of Social Vulnerability in the Province of Salta, Argentina." *Revista de Salud Pública* XVII (2): 46–52. https://pdfs.semanti cscholar.org/a764/0a4ace94b34d6771d11a4e4f38fc38954207.pdf.

Anselin, Luc. 2019. "The Moran Scatterplot as an ESDA Tool to Assess Local Instability in Spatial Association." In *Spatial Analytical*, edited by M. Fischer, H. J. Scholten, and D. Unwin, 111–26. London: Routledge.

Busso, G. 2001. "Vulnerabilidad Social: Nociones e Implicancias de Políticas Para Latinoamerica a Inicios Del Siglo XXI." In *Seminario Internacional Las Diferentes Expresiones de La Vulnerabilidad Social En América Latina y El Caribe.* Chile: Santiago. http://www.redadultosmayores.com.ar/buscador/files/ORGIN011. pdf.

Con, Melina. 2009. "¿Somos Todos Vulnerables? La Vulnerabilidad y Su Heterogeneidad En La Región Metropolitana de Buenos Aires." *Redatam Informa* 15: 5–10. https://repositorio.cepal.org/bitstream/handle/11362/36978/1/RD20 09vol15_es.pdf.

Esperón-Rodríguez, Manuel, Martín Bonifacio-Bautista, and Víctor L. Barradas. 2016. "Socio-Economic Vulnerability to Climate Change in the Central Mountainous Region of Eastern Mexico." *Ambio* 45 (2): 146–60. doi:10.1007/s13280-015-0690-4.

Huffman, Curtis, Paloma Villagómez Ornelas, and Delfino Vargas Chanes. 2019. "La Estructura de Los Hogares y El Ahorro En México: Un Enfoque de Clases Latentes." *Notas de Población* 108: 37–68. https://repositorio.cepal.org/bitst ream/handle/11362/44677/S1900093_Huffman_es.pdf?sequence=1&isAllo wed=y.

Ibarrarán, María E., Elizabeth L. Malone, and Antoinette L. Brenkert. 2009. "Climate Change Vulnerability and Resilience: Current Status and Trends for Mexico". *Environment, Development and Sustainability* 12: 365–388. https://doi.org/10.1 007/s10668-009-9201-8.

Instituto Nacional de Estadística y Geografía. 2015a. *Encuesta Intercensal 2015. Marco Conceptual.* Mexico: INEGI. doi:304.601072.

Instituto Nacional de Estadística y Geografía. 2015b. Encuesta Intercensal 2015. Síntesis Metodológica y Conceptual. Mexico: INEGI. http://www3.inegi.or g.mx/sistemas/biblioteca/ficha.aspx?upc=702825078836.

International Labour Office. 2003. *Working out of Poverty. International Labour Conference. 91st Session 2003. Report of the Director-General\r24413.* Geneva. http: //www.ilo.org/public/english/standards/relm/ilc/ilc91/pdf/rep-i-a.pdf.

Martínez-Rodríguez, Julio C., Néstor R. García-Chong, Laura E. Trujillo-Olivera, and Lucio Noriero-Escalante. 2015. "Inseguridad Alimentaria y Vulnerabilidad Social En Chiapas: El Rostro de La Pobreza." *Nutricion Hospitalaria* 31 (1): 475– 81. doi:10.3305/nh.2015.31.1.7944.

Merrick, Thomas. 2002. "Population and Poverty: New Views on an Old Controversy." *International Family Planning Perspectives* 28 (1): 41–46. https://www.popline. org/node/185689.

Narayan, Deepa. 2002. *Empowerment and Poverty Reduction.* 1st ed. Washington, DC: International Bank for Reconstruction and Development. The World Bank. doi:10.1596/0-8213-5166-4.

Neri, Carolina, and Víctor Magaña. 2016. "Estimation of Vulnerability and Risk to Meteorological Drought in Mexico." *Weather, Climate, and Society* 8 (2): 95–110. doi:10.1175/WCAS-D-15-0005.1.

Pérez-Maqueo, Octavio, M. Luisa Martínez, Flor C. Sánchez-Barradas, and Melanie Kolb. 2018. "Assessing Nature-Based Coastal Protection against Disasters Derived from Extreme Hydrometeorological Events in Mexico." *Sustainability (Switzerland)* 10 (5). doi:10.3390/su10051317.

Pizarro, Roberto. 2001. *La Vulnerabilidad Social y Sus Desafíos: Una Mirada Desde América Latina Estudios Estadísticos y Prospectivos.* Latin American and Caribbean Economic Association. 1st ed. Santiago: ECLAC. http://repositorio.cepal.org/b itstream/handle/11362/4762/S0102116_es.pdf.

Roche, José Manuel. 2008. "Monitoring Inequality among Social Groups: A Methodology Combining Fuzzy Set Theory and Principal Component Analysis1." *Journal of Human Development* 9 (3): 427–52. doi:10.1080/14649880802236706.

Rubio Herrera, Amada, and Fátima Flores Palacios. 2018. "Vulnerabilidad y Su Uso En La Política Social Del Estado de Yucatán. La Dirección de Atención a La Infancia y La Familia." *LiminaR. Estudios Sociales y Humanísticos* 16 (2): 118. doi:10.29043/liminar.v16i2.601.

Ruiz Rivera, Naxhelli. 2012. "La Definición y Medición de La Vulnerabilidad Social. Un Enfoque Normativo." *Investigaciones Geográficas, Boletín Del Instituto de Geografía, UNAM* 77: 63–74. http://www.scielo.org.mx/pdf/igeo/n77/n77 a6.pdf.

Sánchez, Diego, and Carmen Egea. 2011. "Enfoque de Vulnerabilidad Social Para Investigar Las Desventajas Socioambientales. Su Aplicación En El Estudio de Los Adultos Mayores." *Papeles de Poblacion* 17 (69): 1–35.

Sinding, Steven W. 2009. "Population, Poverty and Economic Development." *Philosophical Transactions of the Royal Society of London Series B Biological Sciences* 364 (1532): 3023–30. doi:10.1098/rstb.2009.0145.

Soares, Denise, and Isabel Gutiérrez. 2011. "Vulnerabilidad Social, Institucionalidad y Percepciones Sobre El Cambio Climático: Un Acercamiento Al Municipio de San Felipe, Costa de Yucatán." *CIENCIA Ergo Sum* 18 (3): 249–63. https://ww w.redalyc.org/articulo.oa?id=10420073006.

Soares, Denise, and Daniel Murillo-Licea. 2013. "Gestión de Riesgo de Desastres, Género y Cambio Climático. Percepciones Sociales En Yucatán, México." *Cuadernos de Desarrollo Rural* 10 (72): 181–99.

Stampini, Marco, Marcos Robles, Mayra Sáenz, Pablo Ibarrarán, and Nadin Medellín. 2016. "Poverty, Vulnerability, and the Middle Class in Latin America." *Latin American Economic Review* 25 (1): 4. doi:10.1007/s40503-016-0034-1.

Stern, Claudio. 2012. El "Problema" Del Embarazo En La Adolescencia. Contribuciones a Un Debate. Mexico City: El Colegio de Mexico. http://www.jstor.org/stable/j. ctt14jxqkf.

Vilar-Compte, Mireya, Sebastian Sandoval-Olascoaga, Ana Bernal-Stuart, Sandhya Shimoga, and Arturo Vargas-Bustamante. 2015. "The Impact of the 2008 Financial Crisis on Food Security and Food Expenditures in Mexico: A Disproportionate Effect on the Vulnerable." *Public Health Nutrition* 18 (16): 2934–42. doi:10.1017/S1368980014002493.

Villalobos-Hernández, Aremis, Lourdes Campero, Leticia Suárez-López, Erika E. Atienzo, Fátima Estrada, and Elvia De la Vara-Salazar. 2015. "Embarazo Adolescente y Rezago Educativo: Análisis de Una Encuesta Nacional En México." *Salud Pública de México* 57 (2): 135. doi:10.21149/spm.v57i2.7409.

Zúñiga, F. Ramón, Janette Merlo, and Max Wyss. 2014. "On the Vulnerability of the Indigenous and Low-Income Population of Mexico to Natural Hazards. A Case Study." In *Geoethics: Ethical Challenges and Case Studies in Earth Sciences*, 381–91. Amsterdam: Elsevier. doi:10.1016/b978-0-12-799935-7.00031-9.

7

The Evolving American Opioid Crisis: An Analysis of Gender, Racial Differences, and Spatial Characteristics

Ryan Baxter Hanson and Esra Ozdenerol

CONTENTS

Introduction

The United States is in the midst of an opioid crisis that has developed over the last 30 years. The epidemic is part of a larger trend of drug abuse in which annual rates of drug overdoses have increased exponentially since the 1980s (Jalal et al. 2018). In 2017, 67.8% of drug overdoses were attributed to opioids, which accounted for 47,600 deaths (Scholl 2019). Between 1999 and 2017, there were almost 400,000 opioid-related deaths, and opioid overdose deaths were 6 times higher in 2017 than in 1999 (CDC 2017). Between 2016 and 2017, the unintentional overdose mortality rate involving synthetic opioids rose 45.5% (5.5 to 8.0 deaths per 100,000) (CDC 2017).

The opioid epidemic has a complex demography affected by gender, age, race, urbanicity, the opioid drug in question, historical developments, and location. It cannot easily be designated to one set of demographic components that describe individual victims (Kolodny 2017; Moran 2018; Phillips et al. 2017; James and Jordan 2018; Shihipar 2019; Dasgupta et al. 2018). It is an oversimplification to assign one demographic profile to the epidemic. None the less, media coverage and policymakers have focused on the rise of deaths among male, white, middle-aged, middle-class, rural, and suburban users (James and Jordan 2018; Dasgupta et al. 2018). The epidemic has impacted multiple races in varying locations. Recently, new classes of opioids have begun to cause increases in mortality that have impacted younger groups in more urban environments (Phillips et al. 2017; Scholl 2019).

This chapter examines the gender differences and spatial evolution of the opioid epidemic. The history of the epidemic is explored in relation to the CDC's National Vital Statistics System mortality (NVSS-M) multiple causes of death dataset via the WONDER online database (CDC 2017). The data are used to highlight the impact the epidemic has had in relation to various demographics such as age, gender, and race by location based on varying opioid classifications used in the database.

Classifying Opioids

Opioids are a broad class of drugs that are prescribed for the treatment of pain and can be abused recreationally (Krieger 2018). There are several basic types of opioids, which include natural opioids derived from the resin of the poppy plant, such as morphine and codeine; semi-synthetic opioids such as hydrocodone, oxycodone, or buprenorphine; and fully synthetic opioids which are created in a laboratory and include drugs such as fentanyl and methadone (Opiate Addiction and Treatment Resource 2013). Synthetic opioids can be 50 to 100 times as potent as the natural opioid such as morphine (HHS 2017).

Prescription opioids are those prescribed by a physician for pain management and can be natural, semi-synthetic, or synthetic (Hall et al. 2006). Prescription opioids can also be prescribed for the treatment of opioid addiction (Hall et al. 2006). Methadone and buprenorphine are two opioids used in this way (Hall et al. 2006). Table 7.1 contains a list of the general classifications of opioid drugs with examples of each type.

TABLE 7.1

General Classification of Opioids

Classification	Examples
Natural Opioids	Morphine and Codeine
Semi-synthetic Opioids	Hydrocodone, Oxycodone, and Buprenorphine
Full-Synthetics Opioids	Fentanyl and Methadone
Prescriptions Opioids	Opioid Drugs Obtained with Physician's Prescription
Illicit Opioids	Opioids Obtained without Physician's Prescription

TABLE 7.2

Centers for Disease Control ICD-10 Opioid-Related Classifications

ICD-10 Code	ICD-10 Title	Description	Code Type
X-40–X-44	Drug Poisonings (Overdose) Unintentional	Accidental Overdoses	Underlying Cause of Death
T40.0	Opium	Opium	Multiple Cause of Death
T40.1	Heroin	Heroin	Multiple Cause of Death
T40.2	Other Opioids	Natural and Semi-synthetic Opioids	Multiple Cause of Death
T40.3	Methadone	Methadone	Multiple Cause of Death
T40.4	Other Synthetic Narcotics	Synthetic Opioids	Multiple Cause of Death

In contrast to prescription opioids, illicit opioids are those obtained without a prescribing physician. These can come in many forms, such as the semi-synthetic opioid heroin which is derived from morphine, illegally obtained prescription opioids, counterfeit opioids, and drugs adulterated with synthetic opioids such as fentanyl (Hall et al. 2006).

The CDC classifies opioids under the following International Classification of Diseases, 10th Revision (ICD-10), categories: opium (T40.0), heroin (T40.1), other opioids (T40.2), methadone (T40.3), and other synthetic narcotics (T40.4) (CDC 2018b). The CDC classifies prescription opioid deaths as those caused by natural and other opioids (T40.2) and methadone (T40.3) (CDC 2018b). Table 7.2 contains details of the opioid classifications by the ICD-10.

History of the Opioid Crisis

According to the CDC, the opioid crisis occurred in a series of three waves that were all associated with different classes of opioids (CDC 2018a). The first wave had its roots in the 1980s but started to take shape in the 1990s as physicians increased the prescribing of opioid pain relievers (Kolodny et al. 2015). During the 2000s, the crisis became a public health epidemic as mortality rates began to rise.

The second wave began in 2010 and was associated with a rise in mortality due to the illicit drug heroin (Rudd et al. 2016; Spencer et al. 2019). This was followed by the third wave which saw an increase in mortality from synthetic drugs, such as fentanyl, which were used to adulterate illicit drugs like heroin, counterfeit pills, and cocaine (Scholl 2019; Spencer et al. 2019). Figure 7.1 shows the overall U.S. mortality rate due to opioids, while figure 7.2 shows the mortality rate divided into different classes.

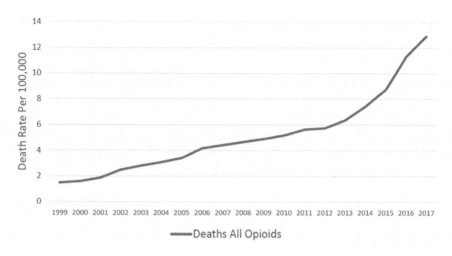

FIGURE 7.1
U.S. annual rate of all opioid overdose mortalities, 1999–2017.

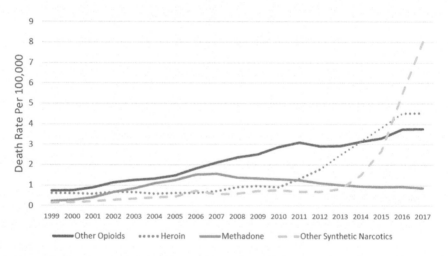

FIGURE 7.2
U.S. annual rate of opioid overdose mortalities by opioid type, 1999–2017.

The first wave of the opioid epidemic has a complex history with multiple causes that include, but are not limited to, questionable academic research, changing practices and opinions on pain management, lobbying by pharmaceutical companies, misinformation from nonprofit organizations backed by the pharmaceutical industry, and aggressive marketing tactics to physicians (Jones et al. 2018; DeShazo et al. 2018; Meldrum 2016). In 1980, Dr. Hershel Jick published a letter in the *New England Journal of Medicine*, after reviewing the cases of 11,882 hospitalized patients who received opioid treatments for pain, in which he concluded that "the development of addiction is rare in medical patients with no history of addiction" (Porter and Jick 1980). The letter became a landmark study that was cited 608 times between 1980 and 2017 (Jones et al. 2018; DeShazo et al. 2018). Additionally, Dr. Kathleen Foley published two articles in 1981 and 1986 that along with Jick's one-paragraph letter became the basis for a 20-year campaign promoting long-term opioid use for the management of chronic, non-cancer-related pain (Meldrum 2016).

Purdue Pharma released the prescription opioid MS Contin in 1984, followed by OxyContin in 1995, which was marketed as a less addictive opioid (DeShazo et al. 2018). The American Pain Association, which received large portions of its funding from Purdue Pharma, proposed in 1995 the concept that pain be measured as the fifth vital sign, an idea which went on to be supported by the Veterans Affairs Medical System, the Joint Commission, the American Medical Association, and the American Academy of Family Physicians (Jones et al. 2018; DeShazo et al. 2018). It was this ideology that spurred physicians to prescribe opioids at increasing rates. Throughout the 2000s, opioid prescription rates and overdose deaths increased. However, this trend began to change in 2010 with the onset of the second wave of the crisis (Hoots et al. 2018). The rate of opioid prescriptions per 100 persons dropped 3.9% annually between 2010 and 2014 and decreased 10.5% annually from 2014 to 2017 (Hoots et al. 2018).

The second wave of the epidemic associated with heroin took effect as mortality rates from prescription opioids leveled off in 2010. Figure 7.2 shows the U.S. annual rate of opioid overdose mortalities by opioid type. One explanation for the reduction in prescription opioid deaths is the introduction of more restrictive prescription drug monitoring programs (PDMP) that limited opioid prescribing and reduced the availability of prescription opioids for misuse and diversion into illicit drug markets (Bachhuber et al. 2019; Strickler et al. 2019; Grecu et al. 2019).

The restrictions placed on prescribing doctors and a lack of prescription drugs available for clandestine use may have led opioid abusers to the cheaper illicit alternative heroin. Prescription opioid abusers have been documented to switch to illicit drugs when prescription opioids are no longer available (NIDA 2018). It is estimated that 4.0–6.0% of people who abuse prescription opioids transition to heroin and that 80.0% of heroin users first abused prescription opioids (NIDA 2018).

However, the correlations between PDMPs and opioid deaths have varied by state. Research which evaluated the impact of PDMPs on mortality found a correlation between implementation and increased mortality from illicit opioids in certain states (Nam et al. 2017). Other research has shown that the effects of PDMPs on opioid mortality have been less conclusive (Fink et al. 2018). The inconclusive effects of PDMPs may be due to differences in the regulatory aspects of each individual state's program. It may also be attributed to differences in the availability of certain drugs among different illicit drug markets (Ciccarone 2017; Carroll et al. 2017).

The third and current wave of the opioid crisis was associated with the adulteration of illicit drugs with synthetic opioids such as fentanyl beginning in 2013. This was associated with increases in heroin use and demand and the introduction of illicitly manufactured fentanyl (Ciccarone 2017). The increased use and demand saw the introduction of heroin being adulterated with fentanyl.

Gender of Opioid Mortality in Relation to Age, Race, and Ethnicity

The CDC's National Vital Statistics System mortality (NVSS-M) multiple causes of death dataset can provide insight into the demography of the opioid epidemic. The CDC's online WONDER database allows users to delineate data by demographics such as age, gender, race, year, location, and urbanicity. Additionally, queries can be filtered to be specific to certain drugs.

The following analysis and figures focus on several types of opioid drugs as defined by ICD-10 classifications. All data were pulled using the underlying cause of death X-40 to X-44, which represents accidental drug overdoses. This excludes drug poisonings that were the result of suicide, homicide, or had an undetermined underlying cause.

Figures that present data for all opioid deaths use ICD-10 codes T.40–T.44. This includes the opioid drug classifications opium, heroin, other opioids, methadone, and other synthetic narcotics. In addition to this combined grouping, three classifications are analyzed individually in this chapter. All data are presented using the most recent year of data available, 2017.

Figures 7.3–7.6 illustrate opioid mortalities from the different opioid drug categories described above for different age groups by gender. A common aspect of the data that is seen in these figures, as well as the figures in the remainder of the chapter, is that the opioid epidemic had a greater impact on males than on females in terms of mortality. This greater rate of male mortality could be based on the tendency for males to be less risk averse (Charness and Gneezy 2012; Pawlowski et al. 2008). Figure 7.3 presents the mortality rates for all opioids by age and gender. Male mortality rates peaked at ages 25–34, with a rate of 38.3 deaths per 100,000 males. This was followed by a steadily decreasing rate for each ten-year age group. This goes somewhat against the idea that the impact was greatest among middle-aged males.

Female mortalities peaked later at ages 35–44 but had a more consistent rate of mortality between the ages of 25–54. This could be due to the lower number of mortalities among females but also shows that mortality among females from all opioids cannot be considered solely a middle-aged phenomenon. The data showed that mortality rates had become unreliable for males

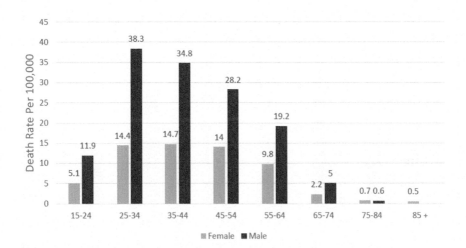

FIGURE 7.3
U.S. rate of all opioid overdose mortalities by age and gender, 2017.

aged 85 and over but were represented for females. This is probably a result of differences in life expectancy between females and males.

Looking at each drug category individually provides further insight into opioid deaths. Figure 7.4 shows heroin-related mortalities for similar age categories as those shown in the previous figure. Heroin mortalities for both genders peaked at ages 25–34, with females at 5.2 and males at 15.3 deaths per 100,000 individuals. Both genders' mortality rates began to decrease with each age group. Like all opioid mortalities, deaths related to heroin decreased with age.

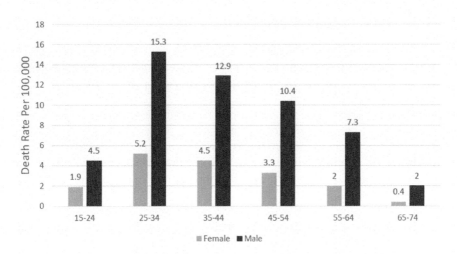

FIGURE 7.4
U.S. rate of heroin overdose mortalities by age and gender, 2017.

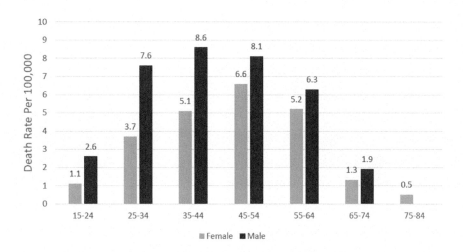

FIGURE 7.5
U.S. rate of other opioid overdose mortalities by age and gender, 2017.

However, data on other opioid mortalities show a stronger pattern of association with middle-aged mortality than did heroin. These mortality rates are illustrated in Figure 7.5. The rates for other opioid deaths were lower than those for heroin in the 15–24 and 25–34 age groups in both genders. However, rates for other opioid deaths among females were higher than for heroin in the age groups between 35 and 64. This shows an association between middle-aged populations and prescription opioid abuse. Perhaps there is a preference for prescription opioids among the middle-aged due to greater access to health care and less access to illicit drug markets among older populations.

In the female age groups of 45–54 and 55–64, the death rates from other opioids were double those for heroin. Male mortality rates from other opioids peaked at ages 35–44, while female mortality rates peaked at ages 45–54. The data suggest that heroin use is more common among younger individuals, particularly males, while other opioid misuse is present in both younger and older populations. Additionally, females are more at risk of mortality from other opioids than from heroin, particularly at ages 35–54. This may be attributed to females having more access to prescription opioids or more opportunities for introduction due to more frequent physician visits. It could also be due to females being more risk averse than are males and perceiving prescription opioids as less dangerous and more socially acceptable drugs than illicit heroin.

The highest rates of mortality linked to an individual opioid drug classification were found in the data for other synthetic narcotics. Synthetic opioids' lethality relates to the strength of synthetic opioids in relation to other opioid classifications and the fact that synthetic opioids such as fentanyl are used as adulterants in heroin and other illicit drugs (HHS 2017; Phillips et al. 2017;

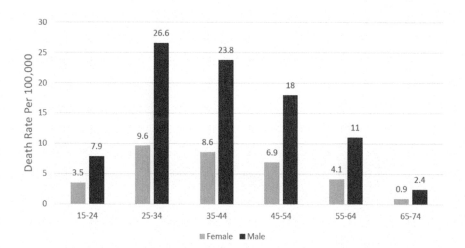

FIGURE 7.6
U.S. rate of other synthetic narcotic overdose mortalities by age and gender, 2017.

Mars et al. 2016). Figure 7.6 shows the mortality rates from other synthetic narcotics by age group and gender. The pattern of mortality among age and gender in the figure mimics the pattern found in the heroin mortalities figure with the exception that the rates are higher. This suggests a correlation between fentanyl's use as an adulterant in heroin and mortalities related to synthetic opioids and heroin. Mortality rates for both genders peaked at ages 25–34 (males at 26.6 and females at 9.6 per 100,000). Like for heroin, the rates taper off with age, suggesting that heroin and illicit drugs adulterated with synthetic opioids are more commonly abused by younger individuals.

It is important to note the differences in mortality rates among females when comparing other opioids and other synthetic narcotics. Females had a higher rate of mortality from other opioids in the 55–64 age group and a similar rate for the 45–54 group. This further shows the association between middle-aged females and prescription opioid abuse. Females are more likely to die from prescription opioids in middle age than from synthetic opioids which are many times more powerful.

Opioid mortalities by drug classification and by race and gender are presented below in figures 7.7 to 7.10. Mortalities for all opioids by race and gender are represented in figure 7.7. White males have the highest death rates from all opioids at 19.9 per 100,000 deaths, followed by African Americans at 16.0, and Native Americans at 11.0. However, there were no significant differences in mortality rates for females from all opioids among whites, African Americans, and American Indian or Alaska Natives. This further demonstrates how males regardless of race have been more greatly affected by the opioid crisis with regard to all opioids. Looking at the data for mortalities due to specific opioid classifications provides further insight into the impact regarding race and gender.

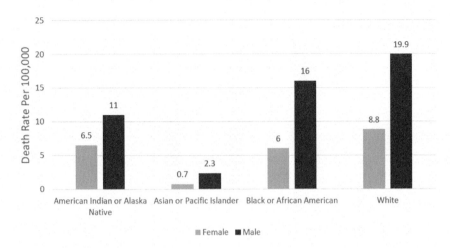

FIGURE 7.7
U.S. rate of all opioid mortalities by race and gender, 2017.

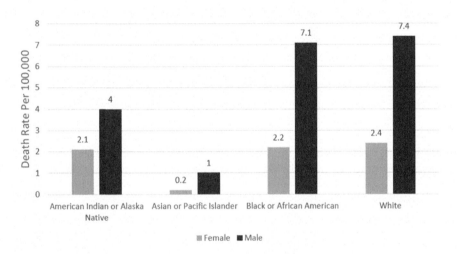

FIGURE 7.8
U.S. rate of heroin overdose mortalities by race and gender, 2017.

The next three figures present the mortality rates by race for individual opioid drugs. Figure 7.8 illustrates the death rates for heroin by race and gender. African American and white males had similarly high mortality rates at 7.1 and 7.4 per 100,000. Asian, African American, and white females all had similar rates of heroin-related mortality (2.1, 2.2, and 2.4 per 100,000). A similar pattern of high rates among white and African American males from other synthetic narcotics is seen in figure 7.9, white males at 12.9 and African American males at 11.7 deaths per 100,000. While female rates from other synthetic narcotics were comparably low for all races. The similar pattern of

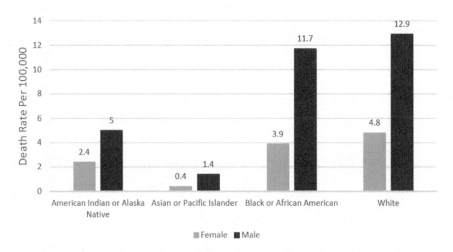

FIGURE 7.9
U.S. rate of other synthetic narcotic overdose mortalities by race and gender, 2017.

mortality between heroin and other synthetic narcotics mirrors the patterns seen for age and mortality shown in figures 7.4 and 7.6. This further supports the association of heroin with synthetic opioid mortality.

Scholars have pointed out that health professionals and the media have falsely portrayed the opioid crisis as a predominately white, male, rural problem which ignored the fact that African Americans have been greatly impacted by the crisis as well (Alexander et al. 2018; James and Jordan 2018; Shihipar 2019). The data support this notion. A longitudinal investigation of the data shows that mortality rates for African American males from heroin have risen along with white male increases (Moran 2018). On top of that, African Americans have been increasingly affected by the epidemic in more recent years. The percentage change of African American mortalities between 2015 and 2017 from all opioids and heroin was more than double that of whites (all opioids, 116.0% and 47.0% change; heroin, 43.0% and 12.0% change). Rates of synthetic opioid mortality for African American males grew 60.0% more than those for white males during the same time period (333.0% and 200.0% change). African American mortality rates have lagged behind white rates but have experienced larger increases in more recent years. This could be due in part to the introduction of fentanyl as an adulter-ant illicit drug.

The data are limited to the time period between 1999 and 2017. It would be beneficial to have rates from earlier periods to see how races were affected differently in the earlier years of the epidemic. It would also be helpful to compare different responses of policymakers to earlier drug epidemics, such as the 1960s heroin and 1980s crack epidemics, which were considered to be associated with African Americans, to the responses to the current opioid epidemic (Cohen 2015; Glanton 2017).

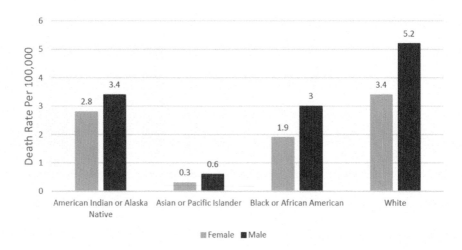

FIGURE 7.10
U.S. rate of other opioid overdose mortalities by race and gender, 2017.

Mortality rates attributed to other opioids by race, seen in figure 7.10, show a different pattern than in the previous figures on race. White males had the highest rate of mortality with respect to both gender and race. Interestingly, the American Indian or Alaska Native race had higher mortality rates than did African Americans. One possible explanation for this could be that physicians are less likely to prescribe opioids to African Americans due to the racist stereotype that they are more likely to misuse or sell the drugs (Lopez 2016; Alexander et al. 2018).

It is important to note that Asian or Pacific Islander mortality rates were low in all opioids and the individual opioid classifications. This could be due to lower rates of drug abuse among Asian races. This can be supported by the data. A search for mortalities using the underlying cause of death codes for accidental overdoses (X-40-X-44) shows that Asian or Pacific Islanders had a much lower rate of mortality (3.1 per 100,000) than did other races regardless of the drugs that caused the mortality (whites, 20.4; Black or African American, 17.8; and American Indian or Alaska Native, 14.1 per 100,000).

In addition to race, the database allows for classification of mortality rates by ethnicity which accounts for two categories, Hispanic or Latino and not Hispanic or Latino. Hispanic or Latino had much lower rates compared to not Hispanic or Latino. The rates were less than half of those for not Hispanic or Latino. However, even among Hispanic or Latino, males have significantly higher rates than do females. Male Hispanic or Latino mortality rates are 2.3–4.5 times higher than for females in studied opioid drug categories. This further shows that opioids had a larger impact on males than on females. Figure 7.11 illustrates these statistics.

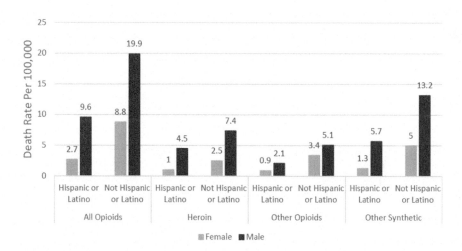

FIGURE 7.11
U.S. rate of all opioid overdose mortalities by opioid type, ethnicity, and gender, 2017.

Urbanicity of Opioid Mortality

In addition to demographics, the CDC's WONDER database allows users to delineate mortalities based on the 2013 Urban-Rural Classification. This classification was created by the National Center for Health Statistics to study health differences among the urban–rural continuum (NCHS 2019). The Urban-Rural Classification consists of six categories of urbanicity at the county level. Urbanicity is determined by whether a county is located within a metropolitan or micropolitan area and the county's population (Ingram and Franco 2014). Table 7.3 contains details about the rules and descriptions of the 2013 Urban-Rural Classification.

Figures 7.12–7.15 show the relationship between opioid overdose mortalities and urbanization classification and gender. Males have higher rates of mortality than do females in all urbanization categories for all opioid drug classifications. Figure 7.12 shows the mortality rates for all opioid drugs. The highest rate for male mortality from all opioids was found in Large Fringe Metros at 21.1 deaths per 100,000, and rates for males were highest in the three more urban categories. The female mortality rates were highest in the Medium Metro categories for all opioid drugs, but unlike the male rates, female rates were more consistent across the urban–rural continuum. The highest rates were among males in urban areas. This is most likely associated with heroin and other synthetic narcotic use. Examples of Large Fringe Metros are counties that tend to be suburban counties of a metropolitan statistical area such as Tipton County in the Memphis, TN-MS-AR statistical area or Dickson County in the Nashville-Davidson-Murfreesboro-Franklin statistical area. Medium Metros are counties also located in metropolitan

TABLE 7.3

2013 Urban-Rural Classification

Level		Urbanization Level	Rule/Description
1	Metropolitan	Large Central Metro	Counties in MSAs of one million or more population that (1) Contain the entire population of the largest principal city of the MSA, (2) Have their entire population contained in the largest principal city of the MSA, or (3) Contain at least 250,000 inhabitants of any principal city of the MSA
2		Large Fringe Metro	Counties in MSAs of 1 million or more population that did not qualify as large central metro counties
3		Medium Metro	Counties in MSAs of populations of 250,000–999,999
4		Small Metro	Counties in MSAs of populations less than 250,000
5	Nonmetropolitan	Micropolitan	Counties in micropolitan statistical areas
6		Noncore	Nonmetropolitan counties that did not qualify as micropolitan

statistical areas but with smaller populations, such as Knox County in the Knox, TN statistical area or Hamilton County in the Chattanooga, TN-GA statistical area.

The mortality rates from heroin and other synthetic narcotics are seen in figures 7.13 and 7.14. Like in previous figures, there seems to be a relationship between these two drugs in that they both had their greatest impact in more

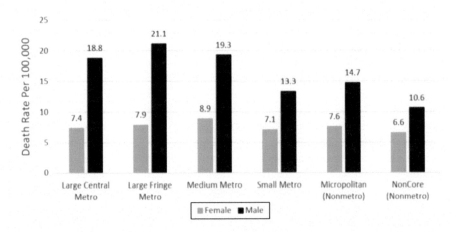

FIGURE 7.12
U.S. rate of all opioid overdose mortalities by 2013 by urbanization and gender, 2017.

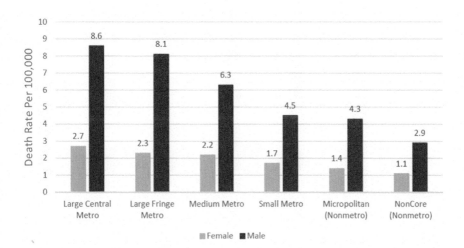

FIGURE 7.13
U.S. rate of heroin overdose mortalities by 2013 by urbanization and gender, 2017.

urban areas. Again, this is most likely due to fentanyl's use as an adulterant of heroin.

The mortality rates associated with heroin decreased for both genders as counties became more rural. However, males were more greatly affected by the drug. Other synthetic narcotic mortality rates showed a similar pattern, having a larger impact on male mortality. Like heroin, synthetic opioids had the highest mortality rates in urban areas. The peak mortality rate for males was 15 deaths per 100,000 in Large Fringe Metros, and the peak for females was 4.9 in Medium Metros. However, there was not the same constant

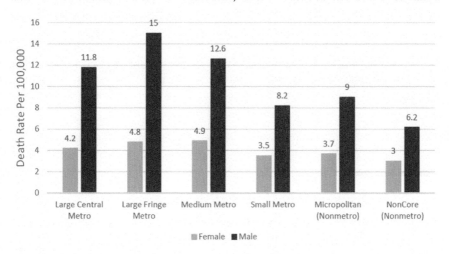

FIGURE 7.14
U.S. rate of other synthetic narcotic overdose mortalities by 2013 by urbanization and gender, 2017.

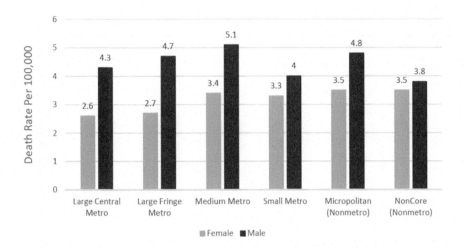

FIGURE 7.15
U.S. rate of other opioid overdose mortalities by 2013 by urbanization and gender, 2017.

decrease of mortality across the urban–rural continuum for other synthetic narcotics that was present in the data for heroin. These two drugs most likely had a larger influence in urbanized areas due to their use as an adulterant in heroin in urban drug markets. The presence of other synthetic narcotics in more rural counties may be associated with fentanyl's use as an adulterant in counterfeit prescription drugs.

Mortality rates associated with other opioids are illustrated in figure 7.15. Males had the highest rates, but the rates for both genders were more random in relation to their urbanicity. There are high rates for both genders in both urban and rural classifications. This goes against the notion of prescription opioids being more abused in suburban or rural settings. The term "hillbilly heroin" has been coined to describe the phenomenon of prescription drugs such as OxyContin being abused by individuals in rural areas due to the lack of accessibility to heroin. Research has shown that this stereotype is not true, and that city dwellers are just as likely to abuse prescription opioids as are individuals living in rural areas (Black and Hendy 2019).

Spatial Aspects of Opioid Mortality

Nationally, opioid overdose mortalities are not evenly distributed, which can be seen in figures 7.16–7.19 (CDC 2017). Figure 7.16 shows the U.S. spatial distribution by state of overdose mortalities due to all opioid drugs for the year 2017. The highest mortality rates were found in the northeastern states of Connecticut, Maine, New Hampshire, and Rhode Island, and in Ohio and

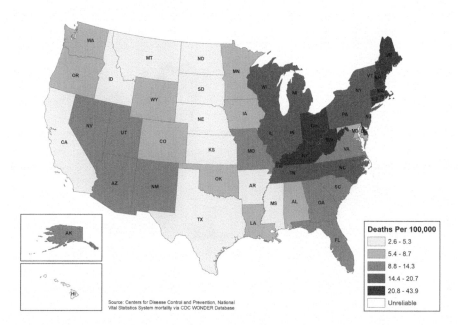

FIGURE 7.16
All opioid overdose deaths per 100,000 by state, 2017.

West Virginia. In general, the highest rates of mortality were found in the Northeast, Midwest, upper Southeast, and New Mexico. The highest rate of mortality due to opioid overdoses was found in West Virginia at 43.9 deaths per 100,000. West Virginia is a rural state. However, many other rural states such as Montana, the Dakotas, and Wyoming had low rates. High rates in the Northeast and Ohio are most likely associated with heroin markets.

The mortality rates from heroin and other synthetic narcotics by state are shown in figures 7.17 and 7.18. Heroin overdose mortalities were concentrated in the midwestern states, northeastern states, and New Mexico. The highest rates were found in the District of Columbia and West Virginia at 17.9 and 13.3 deaths per 100,000, respectively. Synthetic opioid mortality rates were highest east of the Mississippi River in the midwestern and northeastern states. The highest rates were in West Virginia and Ohio at 33.3 and 29.5 deaths per 100,000, respectively. Most researchers believe that other synthetic narcotics' almost exclusive mortality rates east of the Mississippi are due to the difference in heroin drug markets (Mars et al. 2016). West of the Mississippi heroin is supplied from Mexican drug cartels in the form of black tar heroin, while the South American cartels that supply the drug market east of the Mississippi sell more highly processed powder heroin which is more easily adulterated with fentanyl (Mars et al. 2016; Mars et al. 2018). This aspect of the heroin markets has protected the western states from the fentanyl epidemic, but this may change with the

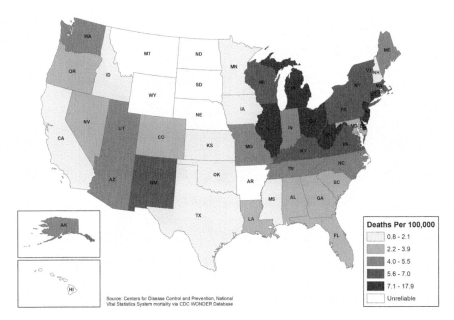

FIGURE 7.17
Heroin overdose deaths per 100,000 by state, 2017.

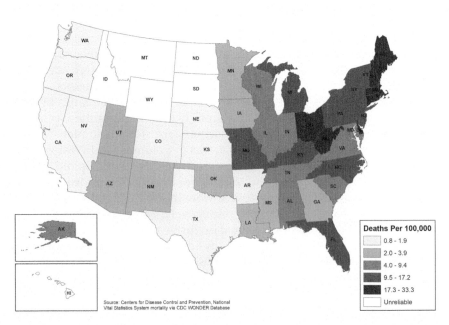

FIGURE 7.18
Other synthetic narcotic overdose deaths per 100,000 by state, 2017.

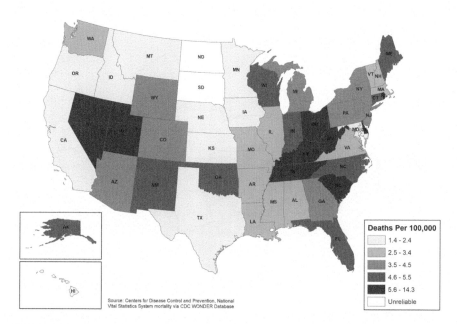

FIGURE 7.19
Other opioid overdose deaths per 100,000 by state, 2017.

reported rise in the use of fentanyl in areas around the Mexican border and as drug suppliers develop ways to adulterate black tar heroin with fentanyl (Sanger-Katz 2018; Debruyne 2019).

Mortalities associated with other opioids had high concentrations throughout the United States. Figure 7.19 shows the distribution of these mortalities. West Virginia had the highest rate of mortality at 14.3 deaths per 100,000. Other areas of high other opioid death rates were located in the Midwest, upper Southeast, Rhode Island, and Delaware on the eastern coast and in the western states of Nevada and Utah. The state of West Virginia had the highest rate of mortality for all the opioid classifications. This may be due to economic factors that affected the state. West Virginia had the second highest annual unemployment rate (5.2%) in the Continental United States behind New Mexico (5.9%), which also had high rates of heroin mortality (BLS 2019).

Utah is another state of interest. Utah ranked relatively high in other opioid mortalities compared to other states. This could be due in part to the state's large membership in the Church of Latter-day Saints. Previous research has shown a correlation between the faith and prescription opioid mortality in western states such as Utah, Idaho, and Wyoming relative to other western states which tend to have overdoses associated with other drugs (Kerry et al. 2016). Members of the church adhered to a stringent health code that prohibited the use of tobacco, alcohol, caffeine, and illicit drugs. It is hypothesized that prescription drugs are viewed as more acceptable for use and thus are more likely to be misused than are illicit drugs.

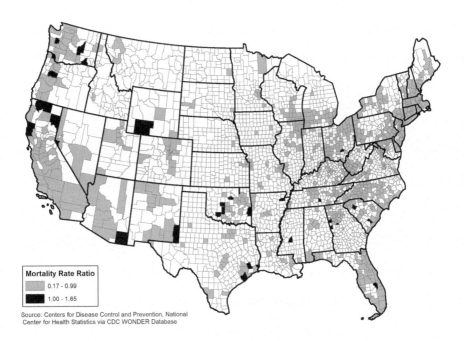

Mortality Rate Ratio
0.17 - 0.99
1.00 - 1.65

Source: Centers for Disease Control and Prevention, National
Center for Health Statistics via CDC WONDER Database

FIGURE 7.20
All opioid mortality rate ratio for males to females by county, 2007–2017.

Male mortality rates were higher than those for females at the state level for all opioid drug classifications in 2017. The exception was for other opioids where females had higher rates than did males in Arkansas, Kansas, Minnesota, and Nevada.

The data were investigated at the county level to further explore the discrepancies between male and female mortality. Rate ratio maps for male-to-female mortality are presented in figures 7.20–7.22. Data for the last decade, 2007–2017, were chosen for use unlike for previous maps which focused on data for 2017. This was done in order to account for small numbers of opioid-related deaths that occurred at the individual county level.

Mortality rates for all opioids seem to be randomly scattered throughout the United States. However, most of the counties where female rates are higher are west of the Mississippi River, with pockets of concentration in the Southwest and the Pacific Coastal regions. A similar pattern is seen with other opioids, but with more dispersion. In Oklahoma, there is an interesting cluster of higher rates for females seen for both all opioids and other opioids.

During the time period considered, no counties had higher female rates of heroin mortality. However, most interesting is the pattern seen with other synthetic narcotics. While previous maps showed that synthetic opioids had had a greater impact east of the Mississippi River at the state level, there is a completely different pattern when looking at the differences between male and female mortality. All counties where women had higher levels of

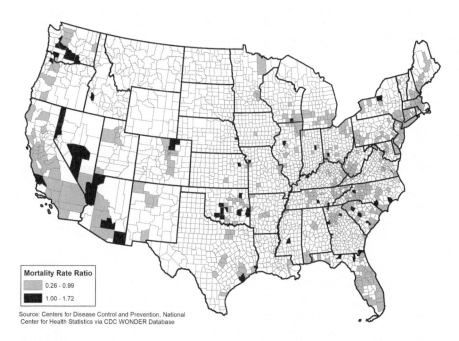

FIGURE 7.21
Other opioid mortality rate ratio for males to females by county, 2007–2017.

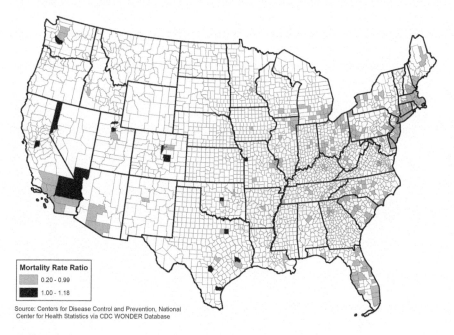

FIGURE 7.22
Other synthetic narcotic mortality rate ratio for males to females by county, 2007–2017.

mortality were west of the Mississippi River. This may be due to lower levels of use in general in these areas.

Conclusions and Discussion

By analyzing the data from the CDC's NVSS-M database on opioid mortality, we found the notion that the opioid crisis had its largest impact on male, white, middle-aged, middle-class, rural, and suburban populations to be an oversimplification of the crisis. One point of this notion which is clear is that the crisis has had a larger impact in terms of male mortality. Males were more likely to die from opioid overdoses than were females. This may have less to do with opioids and more to do with males' overall drug use. According to the data in the CDC's NVSS-M database, it was found that males were more than twice as likely to die from an accidental drug overdose regardless of the drug.

However, this interpretation of the data as it relates to gender may be an oversimplification of the evolving epidemic. Recent research shows that women's heroin use is increasing at a faster rate than that of men, and women's rate of nonmedical use of prescription opioids is reducing more slowly than does males' (Becker and Mazure 2019; Marsh et al. 2018). There are biological differences in how women experience pain, are more greatly affected by opioid drugs, and experience greater withdrawal symptoms (Marsh et al. 2018). In addition, there are non-biological differences such as women experiencing greater mental health issues (Marsh et al. 2018). Another factor that puts women at greater risk is the fact that they are prescribed opioids more frequently, as well as the complications of neonatal drug exposure (Marsh et al. 2018).

The notion that the crisis is associated solely with whites is also incorrect. The data show that African Americans, whites, and Native Americans were all impacted by the crisis. However, this misrepresentation of the epidemic being associated with whites may have been beneficial in influencing how policymakers responded to the crisis. Previous drug epidemics such as the 1960s heroin and 1980s crack cocaine epidemics, which were widely considered associated with African Americans, resulted in the criminalization of drug use and policies such as the War on Drugs (Lopez 2016). By contrast, policymakers and health officials have responded to the current opioid crisis by treating it as a public health threat with strategies such as Narcan distribution, PDMPs, and promoting treatment as opposed to criminalization. This may not have been the case if the opioid crisis were considered an African American problem.

The idea that the crisis has been associated with rural populations may have resulted from the fact that West Virginia was one of the states hardest hit by the opioid epidemic. The state is rural and located in a part of the

Appalachian mountain range that saw economic stagnation in the wake of the collapsing coal industry. The state has become a representation of the epidemic since it was one of the most impacted. However, this fails to describe the epidemic at a national level and in other regions of the United States. When we looked at the data, we found that heroin and fentanyl mortality had greater presences in urban communities, and prescription opioid mortalities were present in both urban and rural counties, debunking the notion of "hillbilly heroin."

The false demographic stereotype also fails to consider the historical development of the epidemic and its three waves. We are currently in the third wave, which is associated with increases in fentanyl mortality. Fentanyl's use is also evolving. Currently, the drug has been more prominent in eastern states due to differences in illicit drug markets. This is changing with increased fentanyl in Mexican drug supplies. Fentanyl is also being used in different ways by abusers as a safer, more reliable, and cheaper alternative to heroin (Szalavitz and Taylor 2018).

Gender and race are important factors to consider when developing health interventions, treatment strategies, and public policy (Becker and Mazure 2019). This is particularly important for an evolving public health threat such as the opioid crisis. Our findings suggest that there should be a critical gender-based approach to treatment and prevention. All data should be reported by gender so that researchers can provide gender-specific treatment and prevention strategies to practitioners and the public. Gender is also an important consideration when formulating drug prescribing practices and policies. A better understanding of the role of gender and race will lead to a more effective response to the current opioid crisis and future drug epidemics.

References

Alexander, M.J., M.V. Kiang, and M. Barbieri. 2018. Trends in Black and White Opioid Mortality in the United States, 1979–2015. *Epidemiology* 29, no. 5 (September): 707.

Bachhuber, M.A., E. Tuazon, M.L. Nolan, H.V. Kunins, and D. Paone. 2019. Impact of a Prescription Drug Monitoring Program Use Mandate on Potentially Problematic Patterns of Opioid Analgesic Prescriptions in New York City. *Pharmacoepidemiology and Drug Safety* 28, no. 5 (May): 734–739.

Becker, J.B., and C.M. Mazure. 2019. The Federal Plan for Health Science and Technology's Response to the Opioid Crisis: Understanding Sex and Gender Differences as Part of the Solution Is Overlooked. *Biology of Sex Differences* 10, no. 1 (January 7): 3.

Black, P., and H.M. Hendy. 2019. Do Painkillers Serve as "Hillbilly Heroin" for Rural Adults with High Levels of Psychosocial Stress? *Journal of Ethnicity in Substance Abuse* 18, no. 2 (June): 224–236.

BLS. 2019. *U.S. Bureau of Labor Statistics.* https://www.bls.gov/.

Carroll, J.J., B.D.L. Marshall, J.D. Rich, and T.C. Green. 2017. Exposure to Fentanyl-Contaminated Heroin and Overdose Risk among Illicit Opioid Users in Rhode Island: A Mixed Methods Study. *International Journal of Drug Policy* 46: 136–145.

CDC. 2017. National Center for Health Statistics via CDC WONDER Database. Multiple Cause of Death Data.

CDC. 2018a. Understanding the Epidemic. Centers for Disease Control and Prevention. Understanding the Epidemic | Drug Overdose | CDC Injury Center. https://www.cdc.gov/drugoverdose/epidemic/index.html.

CDC. 2018b. Opioid Data Analysis and Resources. Centers for Disease Control and Prevention. Opioid Data Analysis and Resources | Drug Overdose | CDC Injury Center. https://www.cdc.gov/drugoverdose/data/analysis.html.

Charness, G., and U. Gneezy. 2012. Strong Evidence for Gender Differences in Risk Taking. *Journal of Economic Behavior & Organization* 83, no. 1. Gender Differences in Risk Aversion and Competition (June 1): 50–58.

Ciccarone, D. 2017. Editorial for "US Heroin in Transition: Supply Changes, Fentanyl Adulteration and Consequences" IJDP Special Section. *International Journal of Drug Policy* 46 (August): 107–111.

Cohen, A. 2015. When Does a Drug Epidemic Get Treated as a Public-Health Problem? *Atlantic*, August 12. https://www.theatlantic.com/politics/archive/2015/08/cr ack-heroin-and-race/401015/.

Dasgupta, N., L. Beletsky, and D. Ciccarone. 2018. Opioid Crisis: No Easy Fix to Its Social and Economic Determinants. *American Journal of Public Health* 108, no. 2 (February): 182–186.

Debruyne, A. 2019. An Invisible Fentanyl Crisis Is Emerging on Mexico's Northern Border. *Pacific Standard*, February 6. https://psmag.com/social-justice/a-fenta nyl-epidemic-is-rising-in-northern-mexico.

DeShazo, R.D., M. Johnson, I. Eriator, and K. Rodenmeyer. 2018. Backstories on the US Opioid Epidemic. Good Intentions Gone Bad, an Industry Gone Rogue, and Watch Dogs Gone to Sleep. *American Journal of Medicine* 131, no. 6 (June): 595–601.

Fink, D.S., J.P. Schleimer, A. Sarvet, K.K. Grover, C. Delcher, A. Castillo-Carniglia, J.H. Kim, et al. 2018. Association Between Prescription Drug Monitoring Programs and Nonfatal and Fatal Drug Overdoses: A Systematic Review. *Annals of Internal Medicine* 168, no. 11 (June 5): 783.

Glanton, D. 2017. Race, the Crack Epidemic and the Effect on Today's Opioid Crisis – Chicago Tribune. *Chicago Tribune*, August 21. https://www.chicagotribune.com /columns/dahleen-glanton/ct-opioid-epidemic-dahleen-glanton-met-201708 15-column.html.

Grecu, A.M., D.M. Dave, and H. Saffer. 2019. Mandatory Access Prescription Drug Monitoring Programs and Prescription Drug Abuse. *Journal of Policy Analysis and Management*: [The Journal of the Association for Public Policy Analysis and Management] 38, no. 1: 181–209.

Hall, W., C. Doran, L. Degenhardt, and D. Shepard. 2006. Illicit Opiate Abuse. In *Disease Control Priorities in Developing Countries*, ed. D.T. Jamison, J.G. Breman, A.R. Measham, G. Alleyne, M. Claeson, D.B. Evans, P. Jha, A. Mills, and P. Musgrove. 2nd ed. Washington, DC: World Bank. http://www.ncbi.nlm.nih.g ov/books/NBK11797/.

HHS. 2017. What Are Opioids? U.S. Department of Health and Human Services. Prevent Opioid Abuse and Addiction | HHS.Gov. https://www.hhs.gov/opioids/prevention/index.html.

Hoots, B.E., L. Xu, N.O. Willson, R.A. Rudd, L. Scholl, K. Mbabazi, L. Schrieber, and P. Seth. 2018. 2018 *Annual Surveillance Report of Drug-Related Risk and Outcome.* Centers for Disease Control and Prevention. https://www.cdc.gov/drugoverdose/pdf/pubs/2018-cdc-drug-surveillance-report.pdf.

Ingram, D., and S. Franco. 2014. NCH Surban–Rural Classification Scheme for Counties. Vital Health Stat 2. National Center for Health Statistics. https://www.cdc.gov/nchs/data/series/sr_02/sr02_166.pdf.

Jalal, H., J.M. Buchanich, M.S. Roberts, L.C. Balmert, K. Zhang, and D.S. Burke. 2018. Changing Dynamics of the Drug Overdose Epidemic in the United States from 1979 through 2016. *Science (New York, N.Y.)* 361, no. 6408.

James, K., and A. Jordan. 2018. The Opioid Crisis in Black Communities. *Journal of Law, Medicine & Ethics* 46, no. 2 (June 1): 404–421.

Jones, G.H., E. Bruera, S. Abdi, and H.M. Kantarjian. 2018. The Opioid Epidemic in the United States-Overview, Origins, and Potential Solutions. *Cancer,* October 9.

Kerry, R., P. Goovaerts, M. Vowles, and B. Ingram. 2016. Spatial Analysis of Drug Poisoning Deaths in the American West, Particularly Utah. *International Journal of Drug Policy* 33 (July): 44–55.

Kolodny, A. 2017. Why Is The Opioid Epidemic Overwhelmingly White? *National Public Radio.* https://www.npr.org/2017/11/04/562137082/why-is-the-opioid-epidemic-overwhelmingly-white.

Kolodny, A., D.T. Courtwright, C.S. Hwang, P. Kreiner, J.L. Eadie, T.W. Clark, and G.C. Alexander. 2015. The Prescription Opioid and Heroin Crisis: A Public Health Approach to an Epidemic of Addiction. *Annual Review of Public Health* 36, no. 1: 559–574.

Krieger, C. 2018. What Are Opioids and Why Are They Dangerous? *Mayo Clinic.* http://www.mayoclinic.org/diseases-conditions/prescription-drug-abuse/expert-answers/what-are-opioids/faq-20381270.

Lopez, G. 2016. Why Are Black Americans Less Affected by the Opioid Epidemic? Racism, Probably. Vox. https://www.vox.com/2016/1/25/10826560/opioid-epidemic-race-black.

Mars, S.G., P. Bourgois, G. Karandinos, F. Montero, and D. Ciccarone. 2016. The Textures of Heroin: User Perspectives on "Black Tar" and Powder Heroin in Two US Cities. *Journal of Psychoactive Drugs* 48, no. 4: 270–278.

Mars, S.G., J. Ondocsin, and D. Ciccarone. 2018. Sold As Heroin: Perceptions and Use of an Evolving Drug in Baltimore, MD. *Journal of Psychoactive Drugs* 50, no. 2: 167–176.

Marsh, J.C., K. Park, Y.-A. Lin, and C. Bersamira. 2018. Gender Differences in Trends for Heroin Use and Nonmedical Prescription Opioid Use, 2007–2014. *Journal of Substance Abuse Treatment* 87: 79–85.

Meldrum, M.L. 2016. The Ongoing Opioid Prescription Epidemic: Historical Context. *American Journal of Public Health* 106, no. 8 (August): 1365–1366.

Moran, M. 2018. How the Opioid Addiction Crisis Was Rendered "White". Psychiatrics News, April 27. World. https://psychnews.psychiatryonline.org/doi/abs/10.1176/appi.pn.2018.5a14.

Nam, Y.H., D.G. Shea, Y. Shi, and J.R. Moran. 2017. State Prescription Drug Monitoring Programs and Fatal Drug Overdoses. *American Journal of Managed Care* 23, no. 5 (May): 297–303.

NCHS. 2019. NCHS Urban-Rural Classification Scheme for Counties. *CDC – National Center for Health Statistics.* https://www.cdc.gov/nchs/data_access/urban_ru ral.htm.

NIDA. 2018. *Opioid Overdose Crisis. Opioid Overdose Crisis | National Institute on Drug Abuse (NIDA).* https://www.drugabuse.gov/drugs-abuse/opioids/opioid-ov erdose-crisis.

Opiate Addiction and Treatment Resource. 2013. *Types of Opioids – Opiate Addiction & Treatment Resource.* http://www.opiateaddictionresource.com/opiates/typ es_of_opioids.

Pawlowski, B., R. Atwal, and R.I.M. Dunbar. 2008. Sex Differences in Everyday Risk-Taking Behavior in Humans. *Evolutionary Psychology* 6, no. 1 (January 1): 147470490800600100.

Phillips, J.K., M.A. Ford, and R.J. Bonnie. 2017. Trends in Opioid Use, Harms, and Treatment. National Academies Press (US). https://www.ncbi.nlm.nih.gov/b ooks/NBK458661/.

Porter, J., and H. Jick. 1980. Addiction Rare in Patients Treated with Narcotics. *New England Journal of Medicine* 302, no. 2 (January 10): 123.

Rudd, R.A., N. Aleshire, J.E. Zibbell, and R.M. Gladden. 2016. Increases in Drug and Opioid Overdose Deaths--United States, 2000–2014. *MMWR. Morbidity and Mortality Weekly Report* 64, no. 50–51 (January 1): 1378–1382.

Sanger-Katz, M. 2018. Bleak New Estimates in Drug Epidemic: A Record 72,000 Overdose Deaths in 2017. *The New York Times*, August 15, sec. The Upshot. https ://www.nytimes.com/2018/08/15/upshot/opioids-overdose-deaths-rising-fe ntanyl.html.

Scholl, L. 2019. Drug and Opioid-Involved Overdose Deaths – United States, 2013– 2017. *MMWR. Morbidity and Mortality Weekly Report* 67. https://www.cdc.gov/ mmwr/volumes/67/wr/mm675152e1.htm.

Shihipar, A. 2019. Opinion | The Opioid Crisis Isn't White. *The New York Times*, February 26, sec. Opinion. https://www.nytimes.com/2019/02/26/opinion/op ioid-crisis-drug-users.html.

Spencer, M.R., M. Warner, B.A. Bastian, J.P. Trinidad, and H. Hedegaard. 2019. Drug Overdose Deaths Involving Fentanyl, 2011–2016. *National Vital Statistics Reports : From the Centers for Disease Control and Prevention, National Center for Health Statistics, National Vital Statistics System* 68, no. 3 (March): 1–19.

Strickler, G.K., K. Zhang, J.F. Halpin, A.S.B. Bohnert, G.T. Baldwin, and P.W. Kreiner. 2019. Effects of Mandatory Prescription Drug Monitoring Program (PDMP) Use Laws on Prescriber Registration and Use and on Risky Prescribing. *Drug and Alcohol Dependence* 199 (June 1): 1–9.

Szalavitz, M., and M. Taylor. 2018. The Radical New Fentanyl Trend That Could Save Lives and Screw Dealers. Vice. https://www.vice.com/en_us/article/wj3nmb/ the-radical-new-fentanyl-trend-that-could-save-lives-and-screw-dealers.

8

How to Undertake an Inequality, Gender and Sustainable Development Analysis: A GIS Approach to Gender Analysis in Pakistan

Ginette Azcona and Antra Bhatt

CONTENTS

Introduction

The 2030 Agenda for Sustainable Development calls for a transformational change to our world where poverty is eradicated, prosperity is sought, and no one is left behind. Five years in, has the world made progress in this shared vision? Assessing progress and gaps across all 17 Sustainable Development

Goals (SDGs) comprehensively, accurately and through an intersectional, gender-aware approach is crucial to answer this question and inform implementation efforts moving forward. Innovative approaches, including the integration of statistical geospatial data with socio-economic information and analysis, is one way in which tracking and monitoring efforts are being significantly enhanced to more effectively guide implementation efforts.

A hallmark of the 2030 Agenda is that it applies to all countries, all people, and all segments of society while promising to address the rights and needs of the most disadvantaged groups as a matter of priority. The "leave no one behind" approach of the 2030 Agenda, specifically aims to identify not only who is left behind but also the ways in which marginalization and exclusion are experienced. Wealth- and income-based discrimination (or class-based discrimination), for example, are common vectors of marginalization and exclusion across countries. Other forms of discrimination are, for example, ethnic disparities or geographic inequalities, particularly in countries that are spatially segregated across ethnic lines. Analyses across countries of who is being left behind demonstrate that among the most disadvantaged are women and girls, who face the compounded effects of gender-based and other forms of discrimination (UN Women 2018).

Gender-based inequalities overlap with other forms of discrimination, leaving women and girls from marginalized groups behind across a range of development indicators. From a measurement perspective, capturing the intricacies of this phenomenon and its impact on the well-being of different groups requires looking closely at the intersection of gender with other forms of discrimination that pushes women and girls from poor and marginalized groups behind. In the Plurinational State of Bolivia, for example, the illiteracy rate varies widely across groups and sub-groups: men overall and women from rich households have low rates of illiteracy. Among women in poor households, however, the illiteracy rate is 23%. The figure increases to 29% for Bolivian women from the Quechua indigenous group (UN Women 2018).

The intersection of gender-based discrimination with inequalities related to wealth, location, and ethnicity "combine to create deep pockets of deprivation across a range of Sustainable Development Goals (SDGs) – from access to education and health care to clean water and decent work". Analysis published by the authors in UN Women 2018 described this phenomenon as "clustered deprivation" – the tendency for deprivations to co-produce and "cluster" together, such that deprivation in one area of well-being is often accompanied by deprivation in another. Using case studies from Pakistan, Nigeria, Colombia, and the United States, the report results pointed to wide inequalities in women's experience across and within countries. In Azcona and Bhatt (2020), the authors elaborate further on the need to undertake an intersectional approach to the measurement of feminist progress.

This chapter builds on this body of work for the Pakistan case study to show where deprivations cluster spatially and the socio-demographic characteristics of the most deprived women and girls in these settings. Going beyond traditional visualization techniques by incorporating a geospatial dimension for the case of Pakistan, reveals that being in a remote rural area or urban slum in Pakistan is associated with poverty, unavailability of clean fuel, unimproved water and/or sanitation facilities, and a lack of access to health facilities. Women and girls from marginalized groups face the greatest deprivations. The multidimensional maps of deprivation derived from this analysis put into sharp focus the tendency of deprivations to spatially cluster together and bolsters the call for an interdependent approach to the 2030 Agenda and its 17 development goals which brings visibility to gender and other forms of discrimination.

Background

The notion that disadvantage is intensified for women and girls living at the intersection of inequalities and discrimination is not new to feminist scholars or human rights experts and advocates. The term "intersectionality" – defined as "the interaction of multiple identities and experiences of exclusion and subordination" – was coined by Crenshaw (1989) to describe the combined effect of gender and race in shaping black women's experiences in the United States. A leading scholar in critical race theory, Crenshaw's description of the phenomenon is palpable: "If you're standing in the path of multiple forms of exclusion, you're likely to get hit by both" – meaning that those doubly disadvantaged based on their gender and race will face deprivations and inequalities that are a direct product of the interaction of the two, a different experience than those experiencing one form of discrimination and not the other. From an analytical perspective, it means that a focus on race without gender is inadequate because the specific deprivations experienced by women within a racial group will go unrecognized (Spierings 2012; EIGE 2019; European Commission 2007).

An intersectional approach is essential for obtaining substantive equality and justice at a systemic level, going beyond individual discrimination, to addressing structural inequalities by questioning and changing legislative and policy frameworks in the longer term (ENAR 2018). It should be noted that while they are related concepts, and often used interchangeably, (additive) multiple discrimination and intersectional discrimination are not the same. On the one hand, (additive) multiple discrimination acknowledges that individuals can experience different types of discrimination, i.e. discrimination on basis of several grounds, operating separately (Ashiagbor 2013). Intersectional discrimination, on the other hand, is "synergistic" (Crenshaw 1989), involving "a combination of various discrimination which, together, produce something unique and distinct from any one form of discrimination standing alone" (Eaton 1994).

Implementing a methodological approach that captures the intersection of different forms of discrimination is not without its challenges. Data limitations, including availability of data on different forms of discrimination, but also data limitations related to sampling and sample size appropriate for capturing sub-groups are one and identifying which forms of discrimination are relevant in each context is another. Wealth and income-based discrimination (or class-based discrimination) as described above are common stratifiers in society , but there are many others that are relevant in some cases/situations and less relevant in others. Geography for example, living in rural areas versus urban areas , can be a strong predictor of well-being in some contexts, and less relevant in others. (UN Women 2018).

To better identify these context-specific factors and to understand for which groups and in what contexts deprivations tend to cluster, spatial-analysis-related innovations, combined with other disaggregation techniques, are crucial. Geographic Information Systems or GIS techniques are one such innovation that allow the mapping of multiple and intersecting group-based inequalities with other forms of deprivations that are spatial in nature. The resulting picture provides insights on differences in well-being across different groups and sub-groups of the population and how these differences play out from a geographic perspective.

In this chapter, we aim to demonstrate how to undertake a Gender, Inequality and Sustainable Development (IGSD) analysis (Azcona and Bhatt 2020), and how GIS information, where available, can provide important insights on the relationship between gender-based deprivations and other forms of inequality related to geographic location and intersection of geography with other group-based inequality. These findings can be vital for informing geographically targeted interventions aimed at reaching the furthest behind. We begin by presenting a brief literature review of gender and GIS in the next section. This is followed by a note on how traditional household survey datasets that also provide spatial information can be used to identify the furthest behind groups of women and girls. Next, the spatial analysis and its findings are presented in the context of Pakistan. The latest 2017 Demographic and Health Survey (DHS) from Pakistan is used for the case study portion of the chapter. The last section concludes with policy implications.

Literature Review

GIS techniques allow researchers to display and analyze information spatially (Staff and Estes 1990) and is increasingly recognized as a tool that can aid researchers in exploring the effects of spatial and geographic circumstances on women's lives (Kwan 2002; McLafferty 2005) and the diverse roles of men and women (Meizen et al. 2012).

Walker and Vajihala (2009) highlight the usefulness of cross-sectoral analysis by merging household Demographic and Health Survey data, community participatory mapping, and spatial transport center data to show that transport barriers, as opposed to unequal decision-making in the home "asking for permission" or knowing the location of health centers, are the most prevalent barriers women face in accessing basic health services in the kingdom of Lesotho. Brown (2003) also explored impacts of road access by overlaying roads in the Yarsha Khola watershed in Nepal with data from a gender-disaggregated resources-use survey administered by the researchers. They found that road access greatly impacted trends in men and women's work lives, with women with road access working longer days than women with less access and men in households with road access completing more "typically" female tasks.

Spatial analysis often builds upon individual-level analyses to explore community-based patterns and assist researchers in calling for geographically targeted interventions. Uthman (2008) used exploratory spatial data analysis (ESDA) of DHS data to display the geographic variation in the age at which women in Nigeria initiate sexual intercourse. Their results suggested that this variation can be attributed to cultural beliefs that encourage early marriage in the regions with low first sexual intercourse initiation ages.

Kibret et al. (2019) used GIS to map DHS data of anemia rates among women of reproductive age at regional, zonal, and cluster levels in Ethiopia. Their findings showed that in addition to individual-level predictors such as limited education, spatial indicators such as living in a rural area and limited access to clean water and improved toilet facilities were associated with being an anemia hotspot. The use of cluster-level data, in place of country or region, allowed the researchers to more precisely identify the five regions in need of more serious interventions to combat high anemia rates. Bearak, Burke, and Jones (2017) also utilize multiple levels of analysis in exploring abortion access in the United States. In moving from a state- to a (smaller) county-level of analysis, the researchers were able to visually show a more precise area in the central United States in which women face the largest disparity in accessing abortion and argue that restrictions imposed at the state-level more negatively affect women living in these rural areas.

Data and Methodology

The IGSD approach focuses on identifying not only *who* is most deprived (or being "left behind" in the language of the 2030 Agenda), but also the multisectoral ways in which marginalization and exclusion are experienced by these groups. The approach is data intensive but important because it makes these groups visible from a statistical standpoint. Our implementation of the approach in past cross-country work reveals not only that those most deprived are often women and girls from marginalized groups, but also that the factors that contribute to their disadvantage do not operate in isolation, they are interrelated and interdependent and thus require integrated interventions and solutions (UN Women 2018; Azcona and Bhatt 2020).

This section illustrates the potential of existing datasets for operationalizing the IGSD approach and capturing spatially segregated socio-economic disadvantage. The GIS modules of Pakistan's 2017 demographic and health survey are used to perform the spatial analysis. Pakistan is chosen as the illustrative case study for several reasons, including sample size, possibility of performing multi-level disaggregation using proxies for income, ethnicity, location, and the availability of a GIS module. The DHS has the added advantage of comprising a broad set of well-being indicators related to poverty, hunger, health, education and so on. In this analysis official SDG indicator or proxy indicators corresponding to 8 of the 17 SDGs are used to measure spatial and socio-economic inequality in well-being. The SDG areas and indicators chosen are the following (a full list of SDG Goals and targets is presented in table 8.1 in the annex):

- SDG 2 (ending hunger): The nutritional status of women aged 18–49 is measured using the body mass index (BMI), where being underweight is defined as having a BMI lower than 18.5 among non-pregnant adult women.

- SDG 3 (health and well-being): Healthcare and decision-making regarding health is measured using the following indicators: "proportion of births not attended by skilled health personnel," "proportion of women and girls stating that they had difficulty in accessing health care facility due to distance," and "percentage of women and girls aged 18–49 who say that they have no say (alone or jointly) in their health care."

- SDG 4 (quality education): The proportion of women and girls with six or less years of education is used to illustrate differences in basic levels of schooling across groups. This proportion is calculated among all women and girls in the sample – which covers those aged 18–49 in Pakistan

- SDG 5 (gender equality): Child marriage rates before the age of 18 are analyzed and compared across different groups of women aged 18–49 to assess progress in eradicating child marriage, a core target of SDG 5.

- SDGs 6 and 7 (clean water, sanitation, and energy): Household access to basic water and sanitation services and use of clean energy for cooking are looked at for these goals. Despite these being household-level indicators, they have important gender implications. Women

TABLE 8.1

Sustainable Development Goals of the 2030 Agenda for Sustainable Development

GOAL
Goal 1. End poverty in all its forms everywhere
Goal 2. End hunger, achieve food security and improved nutrition and promote sustainable agriculture
Goal 3. Ensure healthy lives and promote well-being for all at all ages
Goal 4. Ensure inclusive and equitable quality education and promote lifelong learning opportunities for all
Goal 5. Achieve gender equality and empower all women and girls
Goal 6. Ensure availability and sustainable management of water and sanitation for all
Goal 7. Ensure access to affordable, reliable, sustainable and modern energy for all
Goal 8. Promote sustained, inclusive and sustainable economic growth, full and productive employment and decent work for all
Goal 9. Build resilient infrastructure, promote inclusive and sustainable industrialization and foster innovation
Goal 10. Reduce inequality within and among countries
Goal 11. Make cities and human settlements inclusive, safe, resilient and sustainable
Goal 12. Ensure sustainable consumption and production patterns
Goal 13. Take urgent action to combat climate change and its impacts
Goal 14. Conserve and sustainably use the oceans, seas and marine resources for sustainable development
Goal 15. Protect, restore and promote sustainable use of terrestrial ecosystems, sustainably manage forests, combat desertification, and halt and reverse land degradation and halt biodiversity loss
Goal 16. Promote peaceful and inclusive societies for sustainable development, provide access to justice for all and build effective, accountable and inclusive institutions at all levels
Goal 17. Strengthen the means of implementation and revitalize the Global Partnership for Sustainable Development

Source: UN Global indicator framework for the Sustainable Development Goals and targets of the 2030 Agenda for Sustainable Development

and girls living in households that are deprived in these areas are more likely to face negative health effects and time constraints that limit their opportunities to access education, paid employment, and leisure. These indicators are also important from the spatial deprivations' perspective.

- SDG 8 (decent work and economic growth): The DHS asks respondents aged 18–49 whether they were employed at the time of the survey. This is used as a proxy for labor market participation.

- SDG 11 (sustainable cities and communities): The proportion of women and girls living in households where three or more people share a sleeping room is used as a proxy measure of overcrowding to capture unsatisfied housing needs across groups and subgroups.

Overcrowding in the household is strongly correlated with adverse health effects, including the increased risk of contracting communicable diseases.

The analysis of results are presented in table 8.2 in the annex. These include the national aggregate for each indicator as well as the disaggregation by wealth quintiles and by location (urban and rural).

In addition, differences in outcomes by ethnicities are also presented. The 2017–2018 Pakistan DHS includes a question about the native language which is used to proxy for ethnicity "what is your native language?". The answer is used in the analysis to identify the following five major ethnic groups, for which sufficient sample sizes were available to allow for multi-level disaggregation: Punjabi, Pashtun, Saraiki, Sindhi, and Urdu-speakers (see box 1). This and other stratifiers, including on wealth and location, enable us to undertake a much more granular outlook of the status of women in Pakistan, including further insights into how women's status and well-being differ across groups and sub-groups in Pakistan society (UN Women 2018).

**BOX 1   Characteristics of Ethnic Groups
Captured in the Pakistan Case Study**

Sindhi – Largely concentrated in the poorest and poorer quintiles, live mostly in rural communities, with less than 25% living in cities.

Saraiki – Live mostly in households in the bottom half of the wealth distribution in rural areas.

Punjabi – Largely concentrated in the top half of the wealth distribution; the majority are urban dwellers, but almost 40% live in rural areas.

Pashtun – Distributed across all wealth quintiles, many of them in the middle and poorer groups; they live largely in rural areas.

Urdu-speaking – Live mostly in households in the richest quintile and largely in urban areas.*

*The term "Urdu" in this chapter is used to refer to Urdu-speaking women and girls.

Based on these three dimensions, the location effect, wealth effect and ethnicity effect are calculated. These are defined as the ratios of the most deprived versus least deprived in location, wealth, or ethnicity, respectively, for a specific indicator. Where data allow, differences based on the combination of wealth, location, and ethnicity are analyzed.

TABLE 8.2

Inequalities in SDG-Related Outcomes between different Groups of Women aged 18–49, Pakistan, 2017

Indicator description	Relevant SDG	Poorest rural Saraiki	Poorest rural Sindhi	Poorest rural Pashtun	Poorest rural	Poorest Rural	National aggregate	National Urban	National Richest	Richest urban	Richest urban Punjabi	Richest urban Urdu
Proportion of women aged 18–49, who are underweight (BMI less than 18.5 kg/m²)	SDG 2	25.29	34.48		24.23	23.81	11.36 / 9.05	5.41	3.94	4.56	5.14	4.42
Proportion of women and girls aged 18–49 who do not have an independent/joint say in own health care	SDG 3	50.02	51.87	83.04	55	54.79	53.84 / 49.01	40.79	40.05	38.45	37.7	34.74
Proportion of births not attended by skilled health personnel												

(Continued)

TABLE 8.2 (CONTINUED)

Inequalities in SDG-Related Outcomes between different Groups of Women aged 18–49, Pakistan, 2017

Relevant	Poorest	Poorest rural	Poorest rural	Poorest	Poorest	National	National	National	Richest	Richest	Richest urban	Richest urban
SDG 3 (births in last five years)	60.69	45.52	58.34	54.31	53.96	37.44	30.65	16.17	6.82	7.65	8.87	0.57
Proportion of women and girls aged 18–49 with six or less years of education												
SDG 4	99.1	98.4	99.38	98.07	97.97	79.07	66.81	45.98	23.34	21.84	19.6	9.68
Proportion of women aged 18–49 who were married before age 18												
SDG 5	49.35	52.58	55.5	51.61	52.13	38.81	34.32	26.68	17.53	17.5	14.82	11.19
Proportion of women and girls aged 18–49 with no access to basic drinking water services												
SDG 6	13.12	20.7	39.96	26.24	26.21	15.49	13.86	11.04	8.83	10.1	15.34	3.6

(Continued)

TABLE 8.2 (CONTINUED)

Inequalities in SDG-Related Outcomes between different Groups of Women aged 18–49, Pakistan, 2017

	Relevant	Poorest	Poorest rural	Poorest rural	Poorest	Poorest	National		Richest	Richest urban	Richest urban
Proportion of women and girls aged 18–49 with no access to basic sanitation facilities	SDG 6	62.18	59.78	48.78	52.88	52.64	19.36	10.59	2.09	2.13	1.98	1.55
Proportion of women and girls aged 18–49 with no access to clean cooking fuel	SDG 7	98.35	97.38	97.89	97.94	97.79	50.69	11.5	4.41	0.63	0.06	0.17
Proportion of women aged 18–49 currently not employed	SDG 8	65.98	68.36	94.14	72.22	72.41	82.55	85.54	88.45	88.09	88.69	87.01
Proportion of women and girls aged 18–49 living in												

(Continued)

TABLE 8.2 (CONTINUED)

Inequalities in SDG-Related Outcomes between different Groups of Women aged 18–49, Pakistan, 2017

	Relevant	Poorest	Poorest rural	Poorest rural	Poorest	Poorest	National	National	National	National	Richest	Richest urban	Richest urban
Proportion of women and girls aged 18-49 reporting that they [have] overcrowded housing	SDG 11	94.58	91.63	84.65	90.13	90.3	75.26	72.02	66.51	49.12	50.87	50.16	43.03
cannot access health care due to distance from health facility	SDG 3	66.42	41.66	81.33	61.32	60.14	51.66	41.99	25.53	18.47	16.23	20.32	10.65

Source: Author's calculations based on microdata from National Institute of Population Studies (NIPS) and ICF international 2017

Notes: Select groups are shown given space limitations. Urdu is used as shorthand for Urdu-speaking, see box 1. No access to clean drinking water: Pashtuns reside mostly in the Khyber Pakhtunkhwa region, where reliance on unprotected wells and springs is particularly high. The 2005 earthquake and 2010 floods have further raised concerns about water quality for residents of this region. These and other factors contribute to much higher rates of no access to clean drinking water for Pashtuns overall but especially those from the poorest rural households. Data for BMI is not shown for poorest rural Pashtun due to insufficient sample size. Based on survey characteristics, married is defined as married and/or cohabitation.

Apart from presenting descriptive statistics using traditional visualization methods, this chapter also uses the GIS module of the 2017 Pakistan DHS to spatially map household-level deprivations. These include four dimensions: lack of access to basic water, basic sanitation, usage of unclean fuel, and distance to health facility as a major barrier for accessing health care.

The GIS module of the DHS provides the geographic coordinates of the DHS survey clusters which can be matched with the individual, household, or cluster-level outcomes using GIS platforms such as ArcGIS, QGIS, and/or Tableau. For this chapter, cluster-level aggregates for each indicator were tabulated and matched with the geographic coordinates of the cluster. Additionally, the location, wealth, and ethnic distribution of each survey cluster were also tabulated and mapped. Juxtaposition of the survey clusters deprived in access to basic services and infrastructure with spatial mapping of socio-economic groups reveals how geographic inequalities in well-being mirror ethnic disparities in outcomes, particularly in parts of the country spatially segregated across ethnic lines.

It is important to note that in most DHS household surveys, the sampling clusters are the primary sampling unit, which align with census enumeration areas. However, depending on the scale of the DHS survey, all census enumeration areas may not be covered. . Nevertheless, crucial insights can be revealed from the analysis of this nationally representative survey for Pakistan even if some geographic areas are excluded.

Case Study – Pakistan

Pakistan is home to 207 million people and ranks sixth in the world population (Pakistan Bureau of Statistics 2017). As can be seen in figure 8.1, the country is divided into administrative units (AUs): Balochistan, Islamabad Capital Territory, Khyber Pakhtunkhwa, Punjab, Sindh, Federally Administered Tribal Areas, Azad Jammu and Kashmir, and Gilgit-Baltistan. Punjab, with a total population of 110 million, is the most populated unit, followed by Sindh with a population of 48 million (Pakistan Bureau of Statistics 2017).

Five of Pakistan's eight administrative units (AUs) are mainly mountainous: from west to east, Balochistan, Federally Administered Tribal Areas (FATA), Khyber Pakhtunkhwa (KPK), Azad Jammu and Kashmir (AJK), and Gilgit-Baltistan (GB). The mountain areas are rich in natural resources such as water, forests, pasture, and valuable minerals, including antimony, aragonite (marble), baryte, chromite, fluorite, coal, magnesite, manganese, and soapstone, particularly in KPK, FATA, and Balochistan. However, the country's endowment of rich resources does not necessarily translate into equitable access and benefit sharing for the mountain communities (Golam et al. 2014).

Pakistan is also ethnically diverse. Based on the 2017 census, the Punjabi population comprises 44.2% of the national total. The next largest groups are the Pukhtuns (also known as Pashtuns) at 15.4%, Sindhis at 14.1%, Saraikiat (or Saraiki) at 10.5%, Urdu-speaking Muhajirs at 7.6%, Balochis at 3.6% and

Geograpic Features
— Major Rivers
▨ Mountains

FIGURE 8.1
Map of Pakistan. Source: The DHS Program 2017; Natural Earth 2019; Arzrumtsyan 2015. Note: The designations employed and the presentation of material on the maps do not imply the expression of any opinion whatsoever on the part of the authors and/or their affiliated organizations concerning the legal status of any country, territory, city or area or of its authorities, or concerning the delimitation of its frontiers or boundaries.

other smaller groups making up 4.7% (Pakistan Bureau of Statistics, undated; (UN Women 2018).).

Analysis and Findings

The results indicate that women from marginalized ethnic groups living in poor rural households fare worst across a variety of well-being and empowerment measures (table 8.2). Women and girls from the poorest 20% of households in rural areas fare worse than women and girls from the richest 20% of households in urban areas, across 10 out of 11 dimensions of well-being. Disaggregation by ethnicity revealed further insights: in the case of malnutrition (proxied by a low BMI), Sindhi women and girls from the poorest rural households fare far worse than any other ethnic group across all wealth quintiles and locations. However, the most disadvantaged ethnic group (based on our list of 6 major groups, including "other ethnic minorities") often varies across indicators between the Sindhi, Saraiki, and Pashtun. Amongst women in the poorest rural households, Pashtun women fare worse followed by Saraiki women in 6 of the 11 dimensions of well-being studied. Recent research suggests disadvantages among Pashtun women can be partially attributed to violence and insecurity in the Federally Administered Tribal Areas (FATA) region that has led to a high rates of internal displacement , as well as patriarchal cultural norms written into tribal codes (Mohsin 2013; Kakar 2004). In other dimensions of well-being, such as access to clean cooking fuel, rural poorest households were similarly deprived irrespective of ethnicity.[1]

In Pakistan, the richest women are much more likely to lack employment than the poorest women, 88.5% and 72.4%, respectively. The counterintuitive result is likely related to differences in push and pull factors across social strata. For example, while poverty pushes the poorest women into employment, often precarious, informal and of the unpaid variety, among the richest, significant barriers including biased gender norms, discrimination in wages, and limited job options contribute to low labor force participation rates (Sadaquat 2011).

In this section a short summary of variations in well-being outcomes by wealth, location, ethnicity, and the compounding effect of all three is provided. This is followed by a spotlight on the simultaneously deprived women who face multiple and intersecting inequalities in opportunities in and well-being, as well as in access to basic services and infrastructure.

Wealth Effect

The Demographic and Health Surveys wealth index, "a composite measure of a household's cumulative living standard", is used to indicate economic status. The "poorest" in this analysis refers to the poorest quintile, i.e., women and girls living in households in the poorest 20% of the wealth distribution. The "richest" quintile refers to the wealthiest 20% of households. In 10 of the 11 dimensions of well-being, women and girls from the poorest households are more deprived than those living in the richest households (UN Women 2018). For instance, 52.6% of the women and girls living in the poorest households lack access to basic sanitation facilities versus 2.1% of women and girls living in households in the richest quintile. Hence, the poorest are 25.2 times as likely as the richest to lack access to basic sanitation facilities (figure 8.2).

Location Effect

Deprivations in Rural Areas

Across all 11 dimensions of well-being presented in table 8.2, except for employment, women and girls living in rural households are more likely

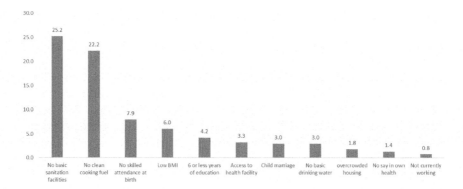

FIGURE 8.2
Wealth effect across dimensions. Source: Author's calculations using National Institute of Population Studies (NIPS) and ICF International 2017.

than those living in urban households to be most deprived. The starkest deprivations are in clean cooking fuel wherein women and girls living in rural areas are 6.4 times as likely to be lacking access to clean cooking fuel when compared to those living in urban areas. That is, while 73.4% of women and girls in rural households lack access, 11.5% of women and girls in urban households do.

Rural women and girls are also about 2 times as likely as to have a lower BMI, to not be attended by a skilled birth attendant, to not have access to basic sanitation facilities and to report difficulty in accessing health care facility due to distance (figure 8.3). Only 3% of women living in rural FATA have completed secondary or higher education, compared to 50% of women living in Islamabad, showing that in this case regional and urban/rural divides overpower gender disparities (NIPS/Pakistan 2019).

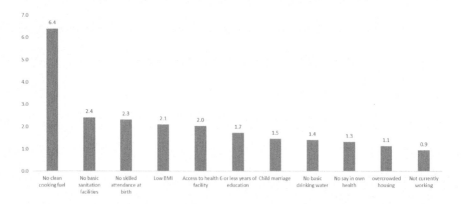

FIGURE 8.3
Location effect across dimensions. Source: Author's calculations using National Institute of Population Studies (NIPS) and ICF International 2017.

Low employment rates among women are widespread in Pakistan, though regional disparities exists: less than 1% of women in FATA report being currently employed at the time of the survey compared to 21% in Sindh and 20% in Punjab (NIPS/Pakistan 2019).

Deprivations in Urban Areas: Slums and Informal Settlements

The world is largely urban and urbanizing further at a great pace. Estimates suggest that by 2050, rapid urbanization in Sub-Saharan Africa and Central and Southern Asia will lead to urban areas accounting for 68% of all people in the world (UN DESA 2018). By 2030, one in every three people will live in a city with at least half a million inhabitants (UN DESA 2016). The rate and scale of urban growth presents many challenges, including the need for investments in transportation, housing, sanitation, energy, and social and physical infrastructure. Where these investments are lacking, a greater number of individuals are forced to live in settlements that lack durable and secure housing, and which are cut off from essential services, such as clean water and sanitation.

Women, facing discrimination in their day-to-day life, with elevated risk of poverty, limited negotiating power and access to resources, are likely to endure the greatest hardships from the proliferation of under-resourced and often spatially segregated urban spaces. As it stands, the latest estimates put the number of people currently living in slum settings at over 1 billion – one sixth of humanity (UN-HABITAT 2016). In developing regions, slum populations are declining as a share of the urban population – from 46.2% in 1990 to 29.7% in 2014 – yet the number of slum dwellers is progressively increasing as a result of urbanization – from 689 million to over 880 million during the same period (UN-HABITAT 2016).

In Pakistan, informal settlements/slum-like conditions[2] are quite prevalent. According to the most recent 2014 estimates, 46% of Pakistan's urban population lives in slum or slum-like conditions (UN-Habitat 2016). Identifying and researching slums has proven historically difficult, given some dispute among researchers and countries on the definition of a slum, as well as the lack of existing relevant and reliable data. Most slum research relies on census data, though it should be noted that census data is often spatially coarse and marked by long gaps between surveys, which is an issue when considering the dynamic nature of slums (Mahabir et al. 2016).

While GIS data mapping using the Demographic and Health Surveys does not allow us to identify all slums or slum-like settlements, a few slum or slum-like survey clusters can be identified. For example, in figure 8.4, a slum or slum-like survey cluster in the Khyber Pakhtunkhwa region is presented, meaning the survey cluster meets the definition of slum settlement based on at least 1 out of the 5 criteria used to define a "slum".

In other words, all individuals living in this survey cluster are deprived in at least one of the following: basic sanitation, basic water, durable housing,

Legend

- ● Large Metropolitan Area
- · Informal Settlements/Slum-Like Conditions

FIGURE 8.4
Slum or slum-like survey cluster in Khyber Pakhtunkhwa. Source: Author's calculations using National Institute of Population Studies (NIPS) and ICF International 2017. Note: SDG indicator 11.1.1 classifies 'slum households' as households that meet at least one of the following five criteria: (1) lack of access to improved water source, (2) lack of access to improved sanitation facilities, (3) lack of sufficient living area, (4) lack of housing durability and (5) lack of security of tenure. These criteria utilize the international definition of 'slum households' as agreed by UN-Habitat, the United Nations Statistics Division and the Cities Alliance. However, in practice, the methodology for measuring security of tenure is not in place; thus, slum status is assessed using the first four criteria only. The designations employed and the presentation of material on the maps do not imply the expression of any opinion whatsoever on the part of the United Nations or the authors concerning the legal status of any country, territory, city or area or of its authorities, or concerning the delimitation of its frontiers or boundaries.

and sufficient living area. Moreover, 83.9% of the people living in this survey cluster do not have access to basic sanitation facilities, and 87.5% do not live in houses with sufficient living area. As is often the case, these slum or slum-like settlements are found next to major cities – in this case Peshawar, one of the megacities in Pakistan. Peshawar is a rapidly growing city due to rural populations and a large number of internally displaced people (IDPs) and Afghan refugees migrating to the city, many of whom arrive to live in slum or slum-like conditions. Past research has shown almost equal sex-ratios in Peshawar's slums (48% females versus 52% males), but high disparities in terms of unemployment, where 96% of women in these slums were unemployed compared to 35% of men (Cynosure Consultants 2013).

In summary, the analysis of least and most deprived, disaggregated by location (rural/urban), reveals that on average rural areas are worst off, but areas of acute deprivation are also observed in urban areas as shown above. Moreover, a thorough look at well-being outcomes in urban areas points to a band of deprivations in slum and slum-like settings.

Compounded Effect

As indicated in UN Women 2018, "the furthest behind are women and girls facing the compounded effect of intersecting forms of discrimination (ethnicity, wealth and location). Ethnicity in some cases exceeds wealth and location as a predicting factor". For example, 34.7% of Urdu-speaking women and girls living in the richest households in urban areas have no say in decisions regarding their own health care (compared to 36.2% for Urdu-speakers overall). However, when compared among Pashtun women and girls, 55% of those in the poorest rural households have no say in their own health care, compared to 76.4% for Pashtun overall and 38.5% in the richest urban households.

Simultaneous Deprivations across Multiple Dimensions of Well-Being by Location, Wealth, and Ethnicity

The compounding effect reveals how different forms of discrimination overlap to leave different groups of women and girls behind in key areas of well-being. However, the IGSD approach goes further by revealing not only 'who' is left behind but also the "how", or rather the multi-sectoral ways in which inequalities manifest. This can be observed by the simultaneous deprivations across multiple dimensions of wellbeing (such as exposure to harmful practices, education, health, and access to remunerated work) that are most acutely felt by women and girls from marginalized and excluded groups. In Pakistan, 13.2% of all women and girls aged 18–49 (or 5.6 million) are simultaneously deprived in four SDG-related dimensions. These women were not only married before the age of 18 and education-poor; they also reported no agency in health care decisions and said they were not working at the time of the survey (figure 8.5).[3]

Women and girls facing simultaneous deprivations across these four dimensions live mostly in rural areas (78.4%) and in households concentrated in the bottom 40% of the wealth distribution (59.6%). The vast majority (72.5%) lack access to clean cooking fuel in their homes. A quarter also lack access to improved sanitation services and 10% reside more than 30 minutes (round trip) from the closest improved water source.

Disaggregation by ethnicity shows Pashtun women to be overly represented among those facing simultaneous forms of deprivation. Sindhi and

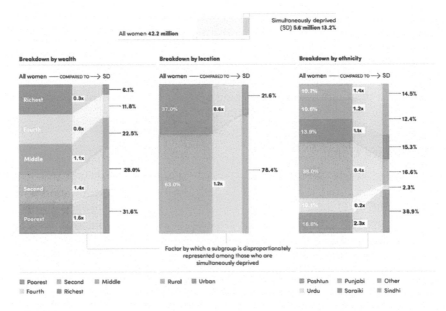

FIGURE 8.5
Proportion of women aged 18–49 married before age 18 and facing education-, employment-, and empowerment-related deprivations, disaggregated by wealth and ethnicity. Source: Author's calculation based on National Institute of Population Studies (NIPS) and ICF International 2017.

Saraiki are also overly represented, as is the "other" category. However, ethnic groups in this "other" category – including Balochi, Barauhi, Hindko, Shina, and many more – are too small to be disaggregated separately.

Multidimensional Deprivation – Spotlight on Additional Simultaneous Deprivations that Are Spatially Distributed

Ethnic disparities mirror geographic inequalities, particularly in countries that are spatially segregated across ethnic lines, geospatial analysis is a useful tool for seeing inequalities. The analysis presented in the section above showed that women and girls from certain ethnic groups were overly represented among those experiencing four simultaneous forms of disaggregation in individual well-being outcomes: these women were not only married before the age of 18 and education-poor, but also reported no agency in health care decisions and said they were not working at the time of the survey. Often these groups who are simultaneously deprived in key individual well-being outcomes are also simultaneously deprived in access to basic services and infrastructure. In this portion of the analysis, using geospatial mapping, we drill down further to the survey cluster level to see how groups facing simultaneous deprivation in well-being are spatially distributed across the country and what additional deprivations they face from the basic services and infrastructure standpoint.

In other words, we posit that women from specific ethnicities such as Sindhi, Pashtun, Saraiki, and "other" ethnic minorities, who face multiple deprivations in individual well-being outcomes also reside in areas that lack access to basic water, sanitation, clean fuel, and report distance to health facility as a major reason in not being able to access a health care facility. This means a further compounding of deprivations and an even more detrimental impact on well-being outcomes.

As can be seen from figure 8.6, Punjabi women and girls are a majority in Pakistan's Punjab region. Juxtaposing this information with figure 8.7, its clear that this densely populated region has the highest proportion of survey clusters with no simultaneously deprived women in access to water, sanitation, fuel, and health services (66%). Similarly, Urdu-speaking women and girls are a majority in clusters in South West Sindh, a part of the country

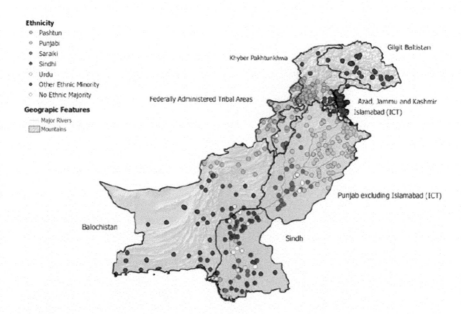

FIGURE 8.6

Survey clusters by major Ethnic groups in Pakistan. Source: Author's calculation based on National Institute of Population Studies (NIPS) and ICF International 2017. *Note*: A survey cluster is defined to have an "Ethnic Majority" if 50% or more of people living in the survey cluster are from a specific ethnicity. "other ethnic minority" category refers to clusters where ethnic minorities such as Balochi, Barauhi, Hindko, Shina, etc. are 50% or more of people living in the survey cluster. The designations employed and the presentation of material on the maps do not imply the expression of any opinion whatsoever on the part of the authors and/or their affiliated organizations concerning the legal status of any country, territory, city or area or of its authorities, or concerning the delimitation of its frontiers or boundaries.

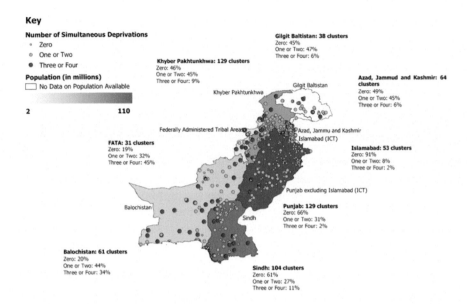

FIGURE 8.7
Survey clusters with simultaneous deprivations. Source: Author's calculation based on National Institute of Population Studies (NIPS) and ICF International 2017. *Note*: The simultaneous deprivations presented in the map are lack of access to basic water, sanitation, availability of clean fuel and distance to health facility as a major reason in not being able to access a health care facility. The designations employed and the presentation of material on the maps do not imply the expression of any opinion whatsoever on the part of the authors and/or their affiliated organizations concerning the legal status of any country, territory, city or area or of its authorities, or concerning the delimitation of its frontiers or boundaries.

with no simultaneously deprived women in access to water, sanitation, fuel, and health services. The opposite pattern is seen for women from Sindhi, Pashtun, Saraiki, and "other" ethnic minorities. For example, Pashtun women and girls are a majority in clusters within the region of FATA and Khyber Pakhtunkhwa. These survey clusters where Pashtun women and girls live (figure 8.6) are the same survey clusters where a majority are simultaneously deprived in three or four dimensions of access to water, sanitation, fuel and health services (figure 8.7).

Cross-country studies have shown that mountainous areas across countries tend to be among the least developed region, presumably because these areas are remote and difficult to get to (Shahbaz et al. 2010). Our survey cluster analysis of Pakistan reveals a similar trend. In the mountainous regions of FATA and Balochistan, a high proportion of survey clusters are classified as having three or four simultaneous deprivations (i.e. majority of women and girls living in this areas face deprivation in at least 3 dimensions), with only 19% and 20% of survey clusters experiencing zero deprivations in each region, respectively. This terrain disproportionately affects Pashtun women and women in "other" ethnic minorities living in these areas. Not having access to basic services as infrastructure, especially in rural areas, means that women and girls are more time-poor given that they disproportionately bear the brunt of unpaid care and domestic work in the household (UN Women 2019).

Conclusion

The 2030 Agenda is heralded as an opportunity to work collectively to transform our world for the better, where the benefits accrue to everyone in society. The monitoring framework that accompanies this promise is critical to delivering real change –it holds leaders accountable to delivering on these commitments. But monitoring progress, particularly on the promise to 'leave no one behind' is not without its challenges. Availability of data and use of statistical methodologies that go beyond summary statistics, such as national average, is paramount to uncover gaps in well-being between groups and sub-groups.

As the case study presented in this chapter shows, overlapping inequalities, for example, those based on gender, ethnicity, geography, and wealth can and often do produce a form of disadvantage that is acute and distinct, leaving women and girls facing these overlapping forms of discrimination worse off than other groups in society, including other groups of women that face gender but not other forms of group-based discrimination. Multi-level disaggregation of data to capture the confluence of gender-based discrimination with other forms of groups based inequality (including ethnicity) brings out these inequalities and is hence critical for identifying the furthest behind.

The analytical procedures used in this chapter show that it is possible to surface the effect of multiple and intersecting forms of discrimination and identify groups of women and girls who experience multiple and simultaneous forms of deprivation in individual well-being outcomes. Triangulating these analytical approaches with GIS data and spatial mapping techniques helps identify the furthest behind by spatially locating groups of women and girls who not only face multiple and simultaneous forms of deprivation in individual well-being outcomes but are also deprived in access to basic services and infrastructure. Doing so represents an important first step that, combined with other sources of information, meaningful social dialogue, and concerted policy action, has the potential to move the promise to leave no one behind from rhetoric to reality. From a statistical perspective, this will require significant investments in collecting, disaggregating, and analyzing data for groups of women and girls who may face multiple and intersecting forms of discrimination.

Acknowledgements

The authors would like to acknowledge Guillem Fortuny Fillo for comments and Julia Brauchle for excellent Research Assistance.

Disclaimer

The chapter significantly builds on the conceptual and empirical work initiated in UN Women 2018 on presentation of multi-dimensional deprivation analysis. Views presented in this chapter are of the authors only and do not represent the views of their affiliated organization. The designations employed and the presentation of material on the maps do not imply the expression of any opinion whatsoever on the part of the authors or their affiliated organizations concerning the legal status of any country, territory, city or area or of its authorities, or concerning the delimitation of its frontiers or boundaries.

Notes

1. The generation of estimates for Urdu-speaking women and girls in the poorest rural households was not possible due to the low sample size (see box on group characteristics above).
2. SDG indicator 11.1.1 classifies "slum household" as households that meet at least one out of five listed criteria: (1) Lack of access to improved water source, (2) Lack of access to improved sanitation facilities, (3) Lack of sufficient living area, (4) Lack of housing durability, and (5) Lack of security of tenure. These criteria utilize the international definition of "slum households" as agreed by UN Habitat-United Nations Statistics Division-UN Cities. In practice, however, methodology and questions for measuring land tenure security (criteria 5) are not in place. Thus, slum status is assessed using the first four criteria only.
3. The case study analysis evaluates well-being in 11 SDG-related dimensions (see table 8.2). Seven of these are at the individual level: BMI, skilled attendance, no say in own health, six years of education or less, child marriage, and employment status. Skilled attendance at birth and BMI, however, were not collected for the full DHS sample and are thus excluded in this portion of the analysis due to sample size constraints. No say in own health care decisions is only collected of women and girls currently married or co-habiting. Therefore, in the case of Pakistan, the clustered analysis refers to sample of women 18-49 currently married or cohabiting at time of survey.

References

Arzrumtsyan, Anna. World Rivers, Distributed by Center for Geographic Analysis at Harvard University, 2015, https://worldmap.harvard.edu/data/geonode:wor ld_rivers_dSe

Ashiagbor, Diamind. *ERA Seminar: Current Reflections on EU Gender Equality Law, The Challenge of Multiple and Intersectional Discrimination in EU Law.* Seminar, University of London, SOAS, April 2013, http://www.era-comm.eu/oldoku/ SNLLaw/11_Multiple_discrimination/2013_04_Ashiagbor_EN.pdf

Azcona, Ginette, and Antra Bhatt. "Inequality, gender, and sustainable development: measuring feminist progress." *Gender & Development* 28, no. 2 (2020): 337–355.

Bearak, Jonathan M., Kristen Lagasse Burke, and Rachel K. Jones. Disparities and change over time in distance women would need to travel to have an abortion in the USA: a spatial analysis. *Lancet Public Health* 2, no. 11 (2017): e493–e500.

Brown, Sandra. Spatial analysis of socioeconomic issues: gender and GIS in Nepal. *Mountain Research and Development* 23, no. 4 (2003): 338–345.

Crenshaw, Kimberle. Demarginalizing the intersection of race and sex: a black feminist critique of antidiscrimination doctrine, feminist theory and antiracist politics. *u. Chi. Legal f.* (1989): 139.

Cynosure Consultants Ltd. *The Study on Slums in Peshawar, Khyber Pakhtunkhwa, Pakistan.* Submitted to UN-Habitat, 2013, http://urbanpolicyunit.gkp.pk/wp-co ntent/uploads/2018/02/Final_Report-_-Study_on_Slums_in_Peshawar_KP_ Pakistan.pdf

Eaton, Mary. Patently confused: complex inequality and Canada v. Mossop. *Review of Constitutional Studies* 1 (1994): 229.

EIGE. *Intersecting Inequalities Gender Equality Index.* Luxembourg: Publications Office of the European Union, 2019, https://eige.europa.eu/publications/intersect ing-inequalities-gender-equality-index

ENAR. European Network Against Racism Symposium. *Intersectionality: A Tool for Equality and Justice in Europe,* 2018, http://www.enar-eu.org/Symposium-In tersectionality-a-tool-for-equality-and-justice-in-Europe

European Commission. Directorate-General for employment, and equal opportunities. Unit G. *Tackling Multiple Discrimination: Practices, Policies and Laws.* Vol. 118. Office for Official Publications of the European Communities, 2007.

Kakar, Palwasha. Tribal law of Pashtunwali and women's legislative authority. *Afghan Legal History Project, Islamic Legal Studies Program, Harvard Law of* School, dátum nem ismert (2004). https://www.semanticscholar.org/paper/Tribal-Law-of-Pas htunwali-and-Women%E2%80%99s-Legislative-Kakar/e6dd9e8624fa9311fe 3569dc069ffe77bdb43c4a.

Kibret, Kelemu Tilahun, Catherine Chojenta, Ellie D'Arcy, and Deborah Loxton. Spatial distribution and determinant factors of anaemia among women of reproductive age in Ethiopia: a multilevel and spatial analysis. *BMJ Open* 9, no. 4 (2019): e027276.

Kwan, Mei-Po. Feminist visualization: re-envisioning GIS as a method in feminist geographic research. *Annals of the Association of American Geographers* 92, no. 4 (2002): 645–661.

Mahabir, Ron, Andrew Crooks, Arie Croitoru, and Peggy Agouris. The study of slums as social and physical constructs: challenges and emerging research opportunities. *Regional Studies, Regional Science* 3, no. 1 (2016): 399–419.

McLafferty, Sara. Women and GIS: geospatial technologies and feminist geographies. *Cartographica: The International Journal for Geographic Information and Geovisualization* 40, no. 4 (2005): 37–45.

Meinzen-Dick, Ruth, B. van Koppen, Julia Behrman, Zhenya Karelina, Vincent Akamandisa, Lesley Hope, and Ben Wielgosz. *Putting Gender on the Map: Methods for Mapping Gendered Farm Management Systems in Sub-Saharan Africa.* IFPRI-Discussion Papers 1153, 2012.

Mohsin, Zakia Rubab. The crisis of internally displaced persons (IDPs) in the federally administered tribal areas of Pakistan and their impact on Pashtun women. *Tigah: A Journal of Peace and Development* 3, no. 2 (2013): 92–117.

National Institute of Population Studies - NIPS/Pakistan and ICF. 2019. *Pakistan Demographic and Health Survey 2017–18.* Islamabad: NIPS and ICF.

Natural Earth. *Cross-blended Hypsometric Tints,* Accessed 11 December 2019. https://www.naturalearthdata.com/downloads/10m-raster-data/10m-cross-blend-hypso/.

Pakistan Bureau of Statistics. 2017. *Provisional Summary Results of the 6th Population and Housing Census 2017.* Accessed 15 December 2017. http://www.pbscensus.gov.pk/.

Pakistan Bureau of Statistics. Undated. *Population by Mother Tongue.* Accessed 15 December 2017. http://pbs.gov.pk/sites/default/files//tables/POPULATIO N%20BY%20MOTHER%20TONGUE.pdf.

Rasul, Golam, A. Hussain, M. A. Khan, F. Ahmad, and A. W. Jasra. *Towards a Framework for Achieving Food Security in the Mountains of Pakistan.* ICIMOD Working Paper 2014/5, 2014.

Sadaquat, Mahpara Begum. Employment situation of women in Pakistan. *International Journal of Social Economics* 38, no 2 (2011): 98–113.

Shahbaz, Babar, Tanvir Ali, Izhar A. Khan, and Munir Ahmad. An analysis of the problems faced by farmers in the mountains of Northwest Pakistan: challenges for agri. extension. *Pakistan Journal of Agricultural Sciences* 47, no. 4 (2010): 417–420.

Spierings, Niels. The inclusion of quantitative techniques and diversity in the mainstream of feminist research. *European Journal of Women's Studies* 19, no. 3 (2012): 331–347.

Staff, J. and Estes, F. *Geographical Information Systems: An Introduction.* Englewood Cliffs, NJ: Prentice Hall, 1990.

The DHS Program. Spatial Data Repository. *Pakistan,* 2017, http://spatialdata.dhspro gram.com/boundaries/#view=table&countryId=PK

UN, DESA. *The World's Cities in 2016 – Data Booklet.* New York: United Nations Department of Economics and Social Affairs, Population Division, 2016.

UN, DESA. *World Urbanization Prospects: The 2018 Revision.* New York: United Nations Department of Economics and Social Affairs, Population Division, 2018.

UN-HABITAT. Slum Almanac 2015–2016: tracking improvement in the lives of slum dwellers. *Participatory Slum Upgrading Programme.* Nairobi: UN-Habitat, (2016).

UN-HABITAT. *Slum Data 2015,* 2016. Accessed 16 December 2019. https://undatac atalog.org/dataset/slum-data-2015.

UN Women. *Turning Promises Into Action: Gender Equality in the 2030 Agenda for Sustainable Development*. New York: UN Women, 2018.
UN Women. *Progress on the Sustainable Development Goals: The Gender Snapshot 2019*. New York: UN Women, 2019.
Uthman, Olalekan A. Geographical variations and contextual effects on age of initiation of sexual intercourse among women in Nigeria: a multilevel and spatial analysis. *International Journal of Health Geographics* 7, no. 1 (2008): 27.
Walker, Wendy, and Shalini P. Vajjhala. Gender and GIS: mapping the links between spatial exclusion, transport access, and the millennium development goals in Lesotho, Ethiopia, and Ghana. Resources for the Future Discussion Paper, no. 09-27 (2009).

9

Visualizing Equality in the New Mobility Workforce

Stephanie Ivey, Tyler Reeb, and Benjamin Olson

CONTENTS

A Geospatial *Tabula Rasa* Moment for Gender Parity?

Transportation systems form the backbone of our economy, connect our communities, and drive societal change. Historically, transportation advances, like the shifts from non-motorized to motorized vehicles or taxis to Transportation Network Companies (TNCs), have dramatically increased not only access and opportunity and led to numerous impacts in myriad sectors as well as societal norms. For example, the advent of the automobile not only provided access to destinations much farther away than ever before but also generated an explosion in related industries (steel, petroleum, etc.), created the need for new infrastructure (such as rest areas, gas stations), and shifted social habits. Similarly, network- and data-driven digital platforms have ushered in a new mobility era where the systems that move people and goods are reshaping individual mobility choices, the economy, and the very notion of what transportation means.

The workforce who will design, develop, operate, and maintain the mobility systems of the future face an era of unprecedented disruption. This trend can be viewed as merely challenging, or as a "tabula rasa" moment to implement workforce priorities that address the most quantitatively significant statistic in transportation workforce development: the gender imbalance. Put plainly, failure to significantly recruit and retain from a population that represents half of the U.S. workforce means millions of critical transportation jobs will go unfilled. Establishing gender parity in the new mobility workforce is essential to filling both traditional and emerging occupations and ensuring that the necessary perspectives required for innovation are present. These priorities are essential in developing a representative and skilled workforce that will quite literally rebuild American transportation systems that serve all U.S. communities.

As advanced technologies continuously evolve and impact the way systems are designed, operated, and maintained, innovations such as Internet of Things (IoT), Artificial Intelligence (AI), and advanced sensors create an environment where autonomous and connected systems are not only possible but probable in the not-so-distant future. These advances fundamentally change the way work is conducted and create new occupations. Historically, much of the transportation operations and maintenance workforce has been comprised of skilled, but largely non-technical, labor. Transformational technologies, however, require these workers to be adept at integration of technology into work tasks, competent with data management and analysis, and flexible and adaptive to continuous learning demands (NNTW, 2019).

In order to navigate a mobility future where algorithms and network technology have as much to do with transportation as roads, rails, and traffic signals, new approaches are needed to synthesize fragmented data into coherent visual narratives that better inform targeted workforce development investments. Fortunately, innovations in Geographic Information Systems (GIS) have created new visual tools to better document and visualize new mobility workforce data. GIS tools make it possible to create data-rich visual narratives that are far more accessible and intuitive than traditional presentation software. Such tools make it possible to turn spreadsheets and lists into visually accessible data-driven frameworks to help policymakers and workforce development leaders address gender parity realities at national, state, and local levels. Said another way, modern GIS makes it possible to develop new visual benchmarks that comprehensively define the new mobility workforce and create a platform for policymakers to better visualize workforce gaps and target investments to address them.

The figures throughout this chapter provide examples of GIS visualizations that begin with national snapshots of gender distribution and progress to state and local data visualizations. Similar visualizations are used in the story map in figure 9.1 by the United Nations that uses data from their Statistics Division to demonstrate disparities in women's economic empowerment across the globe. The visualizations were created as a way to "showcase how the geospatial perspective can enrich the analysis of gender indicators" (United Nations, 2019). The geospatial perspective shown in this story map can be further enriched by observing data at an increasingly localized level as featured in subsequent data visualizations in this chapter.

FIGURE 9.1
United Nations Statistics Division 2017 story map on disparities in women's voice, safety and human rights, and economic empowerment.

Challenges to Attracting and Retaining a Diverse Transportation Workforce

Although the transportation industry is arguably one of the most important underlying elements to both the economy and our communities, societal awareness and perception of job opportunities limit the talent pool (Cronin et al., 2011). Students, parents, and teachers often connect transportation with low-wage, low-skill jobs that have little opportunity for advancement or potential for societal impact. There is limited understanding of the significant connections between transportation careers and science, technology, engineering, and math (STEM). In fact, often the only jobs readily associated with transportation are those of drivers (of trucks, buses, or train engineers). And, there is little understanding of, or respect for, the skill required to be successful in driving roles, and a considerable lack of knowledge about the technologies and operating systems for those vehicles that require drivers to be technically competent. The need to "rebrand" the industry has been proposed by many (Cronin et al., 2011; Ivey et al., 2019; Lockwood and Euler, 2016), but little headway has been made at a national level.

The lack of awareness and rampant misperceptions of the occupations included in the transportation industry are even more concerning when examined in the context of women's participation in the workforce. These misperceptions lead to continued underrepresentation of women in a variety of roles, from operators and technicians to engineers and computer scientists. For example, women account for 7% of truck driving jobs, 20% of engineering technicians, 26% of computer and mathematical occupations, and 14% of civil engineers nationally (BLS, 2019). These broad averages and national statistics, while valuable, do not provide the opportunity to examine patterns in representation across regional boundaries, which could in turn lead to identification of communities where gender parity or near gender parity exists; in those instances best practices can be introduced in other communities. Some of the most important factors determining women's selection of career path include parental input (Rullo, 2019; Ivey et al., 2014), association with societal or community impact (Corbett and Hill, 2015), and role models or mentors (Nathan Associates, Inc., 2017; Rullo, 2019), and these factors should be incorporated into any effort to engage, inspire, and recruit women to pursue careers in the new mobility workforce. Creating new data-rich visual benchmarks makes it possible to give leaders in policy, education, and workforce development a baseline understanding of the state of gender representation in the new mobility workforce. Establishing such a baseline makes it possible to use those benchmarks in longitudinal studies to assess progress in promoting gender parity on local, state, and national levels.

Developing a Robust New Mobility Workforce: The Case for Data and Visualization

Transportation workers include not only vehicle operators but also engineers, computer scientists, IT professionals, planners, and a host of other occupations. Further, technician-level roles, such as drivers, signal/ITS technicians, and other skilled occupations, require that workers are not only familiar with computers and technology but also able to use technology to monitor performance and willing to adapt to continual technological advances (NNTW, 2019). This actuality is much different than the perceptions that are often brought to mind at the mention of "transportation."

One factor contributing to the incomplete picture of transportation held by so many is the lack of data, or lack of data presented in such a way that the diversity of the industry is apparent. For example, data that is readily available via U.S. Bureau of Labor Statistics (BLS) or other national sources designates "Transportation and Material Moving Occupations" or other similar clusters that only highlight one aspect of the industry. Civil engineers (and other STEM professions), for example, are not associated with transportation in such datasets, yet, civil engineers comprise the bulk of the public sector transportation engineering workforce. And, with the rise of new mobility, transformational trends like Smart Cities, Blockchain, Drones, Data Analytics, and Machine Learning are creating jobs that don't have names and job codes yet. Additionally, when looking at broad statistics that are not tied to geography, it is very difficult to tell where pockets of success are occurring, and even more challenging to then examine and link these successes to best practice or investments. This type of aggregation makes it difficult to truly express workforce issues, to isolate factors contributing to both pipeline and diversity challenges, and to tell the story of transportation. This type of data lacks the granular specificity necessary to address gender parity in a targeted and strategic way that helps policymakers effectively invest resources to promote gender parity.

These issues have led to a workforce crisis as the baby boomer generation retires and the new workers required for sustaining the transportation industry are not waiting in the wings (USDOE OCTAE, 2015; Cronin, 2011). As our communities and workforce at large become more diverse (women now comprise 47% of the workforce, and there is projected to be no ethnic majority in the United States by the year 2050) (BLS, 2017a; BLS, 2017b; Pew Research Center, 2015), it is thus imperative for the transportation workforce to follow suit. And, for this to occur, it is necessary to have granular data about specific occupations with specific spatial context. This will allow more robust data analysis, correlation between investments and outcomes, and more meaningful career pathway development.

The new mobility workforce requires gender parity for robust and sustainable progress to be made. This is true not only from a pipeline (number) standpoint but also from an impact on operations, efficiency, and the bottom line. Women must be well represented in transportation occupations to address shortages of talent. Numerous studies have shown that there is a business case for diversity, and in particular a case for women's representation at all levels of an organization (Hunt et al., 2018; McKinsey & Co., 2017; McMahon, 2010).

Gender issues exist in the transportation workforce not only in representation but also in earnings. In order for these disparities to be eliminated, we must first understand at a deep level where inequities are present – requiring data that drills down not only into detailed occupations but also into geography. In many instances, data-rich GIS visualizations can build upon the findings in traditional workforce reports and metrics to provide this additional layer of geospatial analysis. In a landmark effort, the U.S. Departments of Labor, Education, and Transportation released a report in 2015 about workforce development in the transportation industry (U.S. Departments of Labor, Education, and Transportation, 2015). The collaboration was a foundational effort by federal departments that looked at the distribution of attributes such as sex, race, and ethnicity in the transportation workforce, as well as earnings, jBLS, and Economic Modeling Specialists International (EMSI) reports, but the compilation and especially the disaggregation by sector shows how looking at data at a more granular level reveals that certain sectors have more disparity than others.

The report provides data that gives a valuable overview; however, for effective policy and programmatic efforts to be developed, much more in-depth data analysis and visualization of the geographic distribution is required. For example, in Figure 9.2, the 2015 report presents an aggregated look at gender parity, with women representing only 20% of the workforce nationally. While aggregate figures are useful for political talking points and in making headlines, they are often too vague to inform meaningful and targeted policy development. Using GIS visualizations makes it possible to

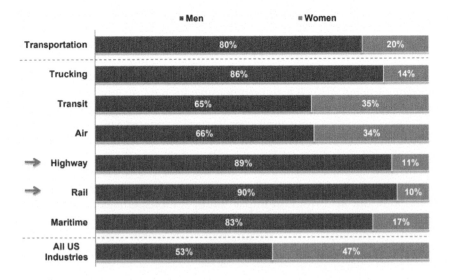

FIGURE 9.2
U.S. Departments of Education, Transportation, and Labor 2014 graph of aggregate transportation workforce figures by subsector.

present more localized realities to provide a more accurate view of gender parity in U.S. communities.

Rather than showing one aggregate percentage for gender representation nationally, figure 9.3 documents that same trend using a state-by-state depiction. That approach makes it possible to compare the percentage of all employed women in each state compared to the percentage of women who are employed in the transportation and material movement (TMM) industry. The comparison in figure 9.3 reveals that even in states with the least amount of gender disparity, there is a maximum of 23% of the transportation workforce comprised of women.

Figure 9.4 further breaks down this data to show the percentage of all employed adults in a given county that are female compared to the percentage of employed adults in the TMM industry that are female for the states of California, Minnesota, and Tennessee. Counties with a similar proportion of

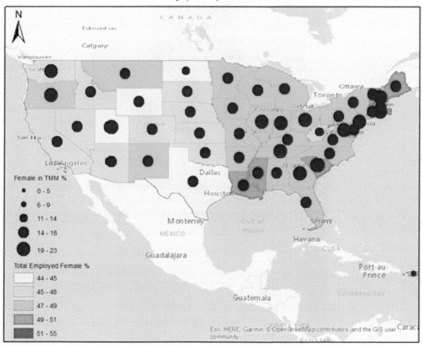

Percentage of Workers in the Transportation and Material Movement Industry (TMM) Who are Female

For Hawaii: Total Employed Female %48. Female in TMM %22
For Alaska: Total Employed Female %46. Female in TMM %15

FIGURE 9.3
Percentage of workers in the transportation and material movement industry (TMM) who are female – United States (created using ArcGIS© software by Esri. ArcGIS© and ArcMap™ are the intellectual property of Esri and are used herein under license. Copyright © Esri. All rights reserved.)

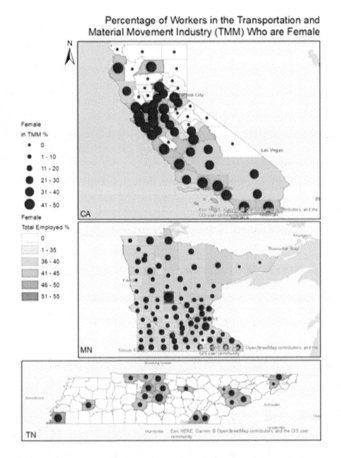

FIGURE 9.4
Percentage of workers in the transportation and material movement industry (TMM) who are female – California, Minnesota, and Florida (created using ArcGIS© software by Esri. ArcGIS© and ArcMap™ are the intellectual property of Esri and are used herein under license. Copyright © Esri. All rights reserved).

each are unsurprising – those with a significant difference can be looked at as case studies to determine what is being done differently in the TMM industry to lead to these differences. For example, a reverse trend in the Dakotas may be due to differences in attraction in some subsectors or could be related to varied industries where STEM workers are employed. Such trends warrant further documentation and analysis. A one-size-fits-all approach may not be appropriate for states where there are inconsistent proportions of females between counties. For states like Tennessee or to a lesser extent California, data are missing for many rural counties and thus a statewide approach is likely to be developed considering only the most populous areas.

Figure 9.5 takes data similar to figure 9.4 but shows that looking at data that is further localized to the city level can reveal differences within

Percent Women in Transportation and Warehousing
Cities in Sacramento County, CA

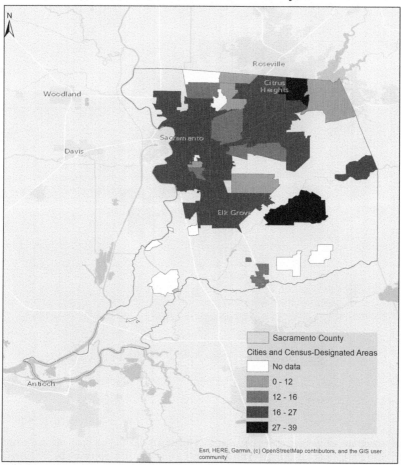

FIGURE 9.5
Percentage of women in transportation and warehousing cities in Sacramento County, CA (created using ArcGIS© software by Esri. ArcGIS© and ArcMap™ are the intellectual property of Esri and are used herein under license. Copyright © Esri. All rights reserved).

counties that need to be considered. Looking at data that is even further localized, such as in Figure 9.6, shows there may be differences within cities that need to be considered as well. This could hypothetically continue down to the single-observation level, which becomes very inefficient, so policy and decision-makers need to decide what granularity is most appropriate for the decision at hand.

GIS visualizations also make it possible to document gender representation according to specific occupations in the transportation workforce. igure

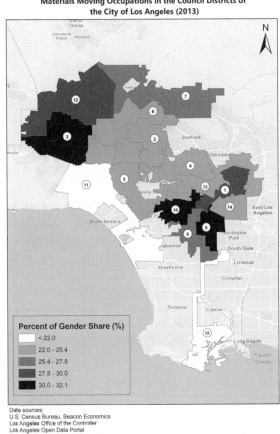

Employment in Production, Transportation, and
Materials Moving Occupations in the Council Districts of
the City of Los Angeles (2013)

Data sources:
U.S. Census Bureau, Beacon Economics
Los Angeles Office of the Controller
Los Angeles Open Data Portal

FIGURE 9.6

Employment in production, transportation, and materials moving occupations in the council districts of the City of Los Angeles (2013) (created using ArcGIS© software by Esri. ArcGIS© and ArcMap™ are the intellectual property of Esri and are used herein under license. Copyright © Esri. All rights reserved).

9.7 shows the percentage of employed adults in each occupation who are female. The distribution of gender for TMM occupations is useful because there are larger disparities in some occupations than in others, as shown in figure 9.7. These disparities are not taken into consideration when looking at the distribution of gender for the industry as a whole – even when looking at more geographically granular data. One complicating factor is that technical, design, and leadership positions in transportation are dominated by STEM professionals. However, in most available data, these occupations are not identified by industry. This makes it difficult to portray this important segment of new mobility occupations. For instance, figure 9.8 shows the percentage of women in STEM occupations by state. It is not possible to

Percent Women in each Occupation 2018
All Transportation and Material Moving Occupations

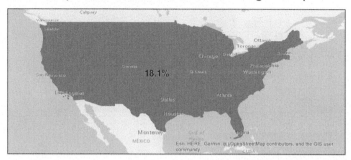

18.1%

Supervisors of TMM Occupations

23.2%

Driver/sales workers and truck drivers

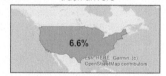

6.6%

Railroad Conductors and Yardmasters

7.1%

Crane and Tower Operators

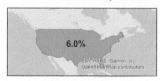

6.0%

Cleaners of Vehicles and Equipment

21.4%

Bus Drivers

43.8%

FIGURE 9.7

Percentage of women in each occupation, 2018 (created using ArcGIS© software by Esri. ArcGIS© and ArcMap™ are the intellectual property of Esri and are used herein under license. Copyright © Esri. All rights reserved).

disaggregate this data to represent STEM professionals in the new mobility industry. Figure 9.9 presents the percentage of women in TMM as compared to women in STEM for each state. This figure demonstrates a lack of coherent connection between these trends and demonstrates the need for additional investigation (and awareness of STEM/transportation relationships). This points to the need for a national conversation to reexamine how transportation careers are portrayed and how workforce data are collected and reported.

Women in Science, Technology, Engineering, and Mathematics (STEM) Occupations, 2013

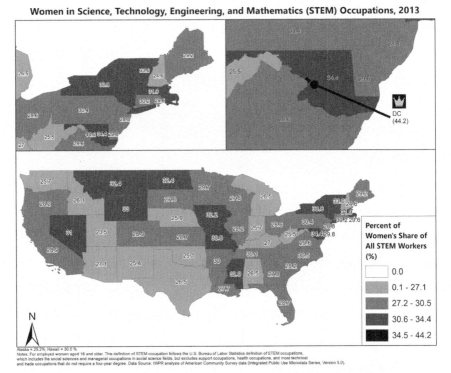

FIGURE 9.8

Women in science, technology, engineering, and mathematics (STEM) occupations, 2013 (created using ArcGIS© software by Esri. ArcGIS© and ArcMap™ are the intellectual property of Esri and are used herein under license. Copyright © Esri. All rights reserved).

One particularly compelling aspect of GIS software is the ability to take data that was previously presented in two-dimensional formats and instead present those same figures in three-dimensional visualizations as depicted in figure 9.10. Three-dimensional visualizations make it possible to more aptly illustrate specific workforce realities. For example, figure 9.10 represents growth in the proportion of females in the transportation workforce with state "height," while shading depicts the overall proportion of females in the workforce. States with higher female representation overall and in the transportation industry, like Rhode Island, can be examined for best practices that might translate to states like Montana. It is also worth noting that interactive digital GIS visualizations can give policymakers deeper insights into issues of gender representation. These tools make it possible to give decision-makers a clean slate or *tabula rasa* sensibility as they consider trends associated with the rapidly evolving new mobility workforce through next-generation visualizations.

Figure 9.11 shows female income disparity at the state and county scale. In the national map, it appears as though there is gender disparity across all states. However, as with most workforce data, when broken down further,

Percent Women in Transportation and Material Moving (TMM) Industry
and in STEM Occupations

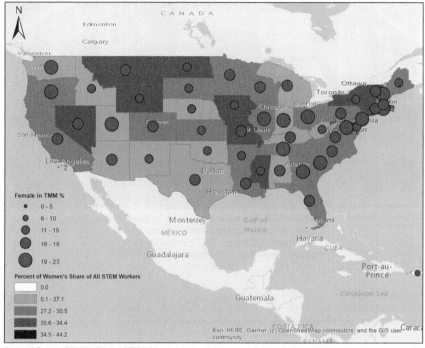

For Hawaii: Female in TMM 22%, Female in STEM 30.0%
For Alaska: Female in TMM 15%, Female in STEM 25.2%

FIGURE 9.9
Percentage of women in transportation and material moving (TMM) industry and in STEM
occupations (created using ArcGIS© software by Esri. ArcGIS© and ArcMap™ are the intellectual property of Esri and are used herein under license. Copyright © Esri. All rights reserved).

there are counties where women are paid more than men. Visualizations like
figure 9.11 make it possible to identify specific communities with gender parity in pay where it may be possible to translate best practices to other communities with imbalances. Figure 9.11 suggests that location plays a large
role in the gender pay gap in transportation. The greatest disparities are seen
in midwestern and southern states, while West Coast and East Coast states
generally show a decreased pay gap between men and women.

Conclusion

The new mobility workforce will increasingly be redefined as transformational technologies and consumer attitudes drive change in personal vehicle,

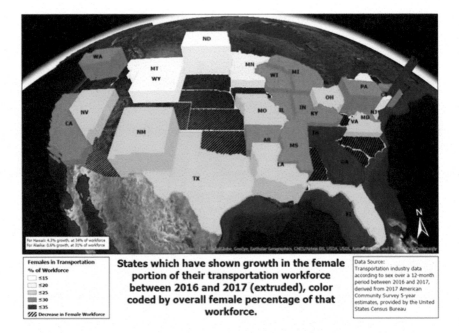

	States which have shown growth in the female portion of their transportation workforce between 2016 and 2017 (extruded), color coded by overall female percentage of that workforce.	
Females in Transportation **% of Workforce** ☐ ≤15 ☐ ≤20 ☐ ≤25 ■ ≤30 ■ ≤35 ▨ Decrease in Female Workforce		Data Source: Transportation industry data according to sex over a 12-month period between 2016 and 2017, derived from 2017 American Community Survey 5-year estimates, provided by the United States Census Bureau

FIGURE 9.10

States which have shown growth in the female portion of their transportation workforce between 2016 and 2017 (extruded), color-coded by overall female percentage of that workforce (created using ArcGIS© software by Esri. ArcGIS© and ArcMap™ are the intellectual property of Esri and are used herein under license. Copyright © Esri. All rights reserved).

mass transit, active transportation, and goods movement, both domestically and internationally. These transformational technologies and trends are, in turn, redefining career pathways for the men and women who comprise the new mobility workforce. This rapid rate of change has ushered in a tabula rasa moment that leaders in industry, education, and government must seize to create new learning models and types of curriculum needed to prepare emerging professionals to develop and operate new transportation systems. These changes also create an environment for attracting and retaining diversity in the industry, but greater efforts must be made to ensure awareness of the future of STEM occupations in the transportation industry. To realize this future, workforce development researchers need to help leaders better understand the changes and workforce disparities using new data visualization techniques.

Indeed, GIS data visualization technologies hold great promise in documenting comprehensive and nuanced data about a wide range of important social issues. Such visualizations will become more compelling and comprehensive as new data trends are documented and, equally important, as new methods of gathering data are developed. Such trends will make it easier to document the state of gender parity in the new mobility workforce.

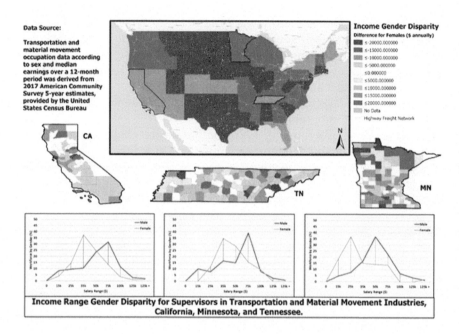

Income Range Gender Disparity for Supervisors in Transportation and Material Movement Industries, California, Minnesota, and Tennessee.

FIGURE 9.11

Income range gender disparity for supervisors in transportation and material movement industries (created using ArcGIS© software by Esri. ArcGIS© and ArcMap™ are the intellectual property of Esri and are used herein under license. Copyright © Esri. Alright rights reserved).

At present, conducting labor market research and analysis on the new mobility workforce is not an easy task. Knowledge, skills, and abilities (KSAs) are redefining existing occupations and entirely new occupations are being created to meet future workforce demands. The BLS has an established repository of Standard Occupational Codes (SOCs) to categorize jobs across major transportation sectors. Given the transformational changes in new mobility, it is not possible to find assigned SOCs for emerging occupations. Further complicating matters is that underlying information and job duties under some existing SOCs are misleading. For example, electric bus technicians are often categorized under diesel bus technician categories because a more appropriate SOC has not been developed. There are many datasets at the national level from the census, BLS, or the Bureau of Transportation Statistics (BTS), but transportation jobs are often categorized as only operators and other low-level positions. Similar to the miscategorization of occupations by SOC codes, excluding critical aspects of the industry does not allow for a comprehensive analysis of gender disparity.

Effective GIS visualizations require robust and accurate data sources. To best look at the transportation workforce gender disparity issue, data gathering methods and online portals have to address the need for more granular, localized data. Current data sharing practices allow for the use of workforce data down to the city or county level, but data are typically not disaggregated beyond the overall gender distribution of the industry in that area. Sector, occupation, or competency-level data are needed to best address the problem. As demonstrated in figures 9.10 and 9.11, GIS visualizations make it possible to conduct analysis at finer scales to present gender disparities in more localized and comprehensive ways to better explain workforce realities and related policy implications.

Data at the state level is readily available, but similar challenges associated with categorization and gender distribution at the occupational level exist. At the local level, general gender distribution data are often available for more populous counties or cities; however, often such data are lacking for rural locations. This leads to policy or industry solutions that fail to consider the entire population of a region. Such data gaps and categorical challenges call for new data collection methods and revised methods for classifying occupations. Herein lies the call for action for promoting gender parity in the new mobility workforce: leaders in industry, government, and education must rethink what data should be collected, how it should be aggregated, and what tools should be used for analysis. With relatively rapid occupational evolutions being driven by advancing technologies, a more flexible framework (and less rigid job categories) is now needed.

In an ideal world, leaders in industry, education, and the U.S. Departments of Labor, Education, and Transportation would have the resources and expertise to update workforce data in real time while accounting for new transdisciplinary trends that are blending entire sectors of the economy and redefining the very notion of what transportation means. However, even with such innovations in data and GIS visualizations, a major factor in determining gender parity remains unaddressed: assessing the representation of women in STEM and leadership positions in transportation. Further, granular data on ethnicity or race of female workers in transportation (by occupation) is needed to provide better understanding of gender parity issues and allow robust analyses to be developed that can point to potential solutions. Another area where additional data is needed is in the documentation by occupation of the age of workers so that pay gaps can also be examined to determine if age is also a contributing factor. Data for other factors such as family status and number of children could also contribute greater understanding regarding specific occupations and implications for career choice. Future research could leverage this data not only for computing basic descriptive statistics but also for integrating spatial analytical methods, correlation analyses and location quotients by gender by industry. For example, using ArcGIS, maps could be created for location quotient values that provide more in-depth perspectives into the varied gender gaps in by industry type.

As of now, gendered data is available in bits and pieces and is housed in a range of public- and private-sector repositories, which makes it cumbersome to do holistic analyses. The visualizations in this chapter provide insight and analysis into data trends associated with a range of transportation occupations,

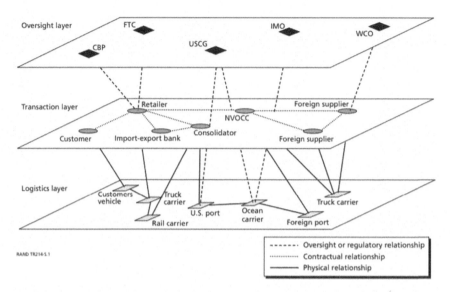

FIGURE 9.12
RAND Corporation 2004 visualization of the supply chain from "Evaluating the Security of the Global Containerized Supply Chain" technical report.

but data on STEM and leadership positions are not available by industry. This lack of data suggests a need for comprehensive data gathering and analysis on STEM occupations as well as methods to more directly gauge the representation of women in leadership positions. Such analysis requires a layered systems approach to documenting and visualizing leadership positions at oversight, transactional, and operational layers as depicted in the modified supply chain model in figure 9.12. Only after accurate and comprehensive workforce data is gathered, analyzed, and made visually accessible at every level – from entry level to the C-suite – will an accurate assessment of gender parity and who controls the levers of power in the new mobility workforce be possible.

The methods of data visualization used in the figures in this chapter suggest ways to document gender disparities to help leaders in industry, government, and education form crucial policy-based measures that attract talented and skilled women to historically male-dominated industries. Such policy measures could help narrow the gender gap and contribute to a more progressive society.

Changing the conversation (as well as attitudes) about transportation careers and gender representation means considering a new way of describing the industry as a whole that is comprehensive, inclusive, and reflective of the innovation and advanced technologies that dominate the sector. GIS visualizations hold great promise in promoting the *tablula rasa* or clean-slate mentality required for leaders in industry, government, and education to understand and empower women to join the ranks of the new mobility workforce. Such empowerment calls for accurate data to identify the right solutions for driving gender parity with targeted investments of public- and private-sector capital to support increased deployments of apprenticeships, internships, and a range of other workforce development programs at both the high school and college levels. These experiential learning opportunities not only improve students' skillsets and understanding of career pathways but also may lead to the development of an "identity" associated with the career, which is a crucial aspect to recruitment and retention of women in the effort to attain gender parity.

Works Cited

"APEC Women in Transportation Data Framework and Best Practices Report". *Nathan*, March 7, 2018. https://www.nathaninc.com/insight/apec-women-in-transportation-data-framework-and-best-practices-report/.

Corbett, Christianne, and Catherine Hill. *Solving the Equation: The Variables for Women's Success in Mathematics and Computing*. Washington, DC: AAUW, 2015.

Cronin, Brian E. *Strategies to Attract and Retain a Capable Transportation Workforce*. Washington, DC: Transportation Research Board, 2011.

"Employed Persons by Detailed Occupation, Sex, Race, and Hispanic or Latino Ethnicity". *U.S. Bureau of Labor Statistics*. U.S. Department of Labor, January 18, 2019. https://www.bls.gov/cps/cpsaat11.htm.

"Employment by Occupation by Gender for Council Districts: Control Panel LA". *Ron Galperin LA Controller*. Los Angeles City Controller, 2019. https://controllerdata.lacity.org/dataset/Employment-by-Occupation-by-Gender-for-Council-Dis/43p4-p3mh.

Hunt, Vivian, Sara Prince, Sundiatu Dixon-Fyle, and Lareina Yee. *Delivering through Diversity*. McKinsey & Company, January 2018. https://www.mckinsey.com/business-functions/organization/our-insights/delivering-through-diversity.

Ivey, Stephanie S., Mihalis M. Golias, Paul Palazolo, Kelsey Ford, Virginia Anne Wise, and Patrice Thomas. Transportation Engineering Careers. *Transportation Research Record: Journal of the Transportation Research Board* 2414, no. 1 (2014): 45–51. https://doi.org/10.3141/2414-06.

Ivey, S. (2019). Inspiring the Next Generation Mobility Workforce through Innovative Industry-Academia Partnerships. In T. Reeb (Ed.), *Empowering the New Mobility Workforce*. p. 317–348, Elsevier, https://doi.org/10.1016/B978-0-12-816088-6.00015-8.

Krivkovich, Alexis, and Irina Starikova. *Women in the Workplace 2017*. McKinsey & Company, 2017. https://www.mckinsey.com/featured-insights/gender-equality/women-in-the-workplace-2017.

"Labor Force Characteristics by Race and Ethnicity, 2016: BLS Reports". *U.S. Bureau of Labor Statistics*. U.S. Department of Labor, October 1, 2017. https://www.bls.gov/opub/reports/race-and-ethnicity/2016/home.htm.

Lockwood, Steve, and Gary Euler. Recruitment, Retention and Career Development. *Transportation System Management & Operations (TSM&O) Workforce Development* 3, June 23, 2016.

Mcmahon, Anne M. Does Workplace Diversity Matter? A Survey Of Empirical Studies On Diversity And Firm Performance, 2000–09. *Journal of Diversity Management* 5, no. 2 (April 2011). https://doi.org/10.19030/jdm.v5i2.808.

"Modern Immigration Wave Brings 59 Million to U.S." *Pew Research Center's Hispanic Trends Project.* Pew Research Center, September 28, 2015. https://www.pew research.org/hispanic/2015/09/28/modern-immigration-wave-brings-59-mil lion-to-u-s-driving-population-growth-and-change-through-2065/.

National Network for the Transportation Workforce. National Transportation Career Pathway Initiative. *National Network for the Transportation Workforce.* Federal Highway Administration, 2019. https://nntw.org/career-pathways.

Rullo, W. S. Impacting Career Choice. In *Career Counseling Across the Lifespan: Community, School, Higher Education, and Beyond,* edited by G. Eliason, J. Patrick, J. Samide, and T. Eliason, 415–446. Charolette, NC: Information Age Publishing, 2019.

"Strengthening Skills Training and Career Pathways across the Transportation Industry". *Strengthening Skills Training and Career Pathways across the Transportation Industry.* Washington, DC: U.S. Department of Education, Office of Career, Technical, and Adult Education, 2015. http://cte.ed.gov/initiatives/ advancing-cte-in-state-and-local-career-pathways-system.

Szymkowski, Todd, Stephanie Ivey, Alexandra Lopez, Pat Noyes, Nicholas Kehoe, and Carrie Redden. Transportation Systems Management and Operations (TSMO) Workforce Guidebook Final Guidebook. *Transportation Systems Management and Operations (TSMO) Workforce Guidebook Final Guidebook.* Transportation Research Board, 2019.

"Trends in Employment and Earnings". *Status of Women in the States.* Institute for Women's Policy Research, July 27, 2015. https://statusofwomendata.org/explo re-the-data/employment-and-earnings/#section-c.

United States Census Bureau, and Department of Commerce. *2017 American Community Survey, 5-Year Estimates,* 2017.

Willis, Henry H., and David S. Ortiz. Evaluating the Security of the Global Containerized Supply Chain. *Evaluation the Security of the Global Containerized Supply Chain.* Rand Corporation, 2004.

"Women and Sustainable Development: Building a Better Future for All..." *United Nations Statistics Division.* United Nations. Accessed July 31, 2019. https://ma ps4stats.maps.arcgis.com/apps/MapSeries/index.html?appid=6bf7ee9779984 6ec950b6ef5c521503d.

"Women in the Labor Force: a Databook: BLS Reports". *U.S. Bureau of Labor Statistics.* U.S. Department of Labor, April 1, 2017. https://www.bls.gov/opub/reports/wo mens-databook/2016/home.htm.

10

Spatial-Temporal Patterns of Gender Inequalities in University Enrollment in Nigeria: 2005–2015

**Moses O. Olawole, Akanni I. Akinyemi, David O. Baloye,
Adesina A. Akinjokun, and Olayinka A. Ajala**

CONTENTS

Introduction

Education is a critical indicator of social and economic well-being. It also enhances individual, family, and community status. Besides, it is an economic indicator that enhances an individual's ability to create wealth, enjoy healthy life, and improve self-actualization of potentials (Kimoso et al. 2015).

Among one of the United Nation's Sustainable Development Goals (SDGs) unveiled as part of the 2030 Sustainable Development Agenda, and adopted by member states in 2015 at a historic UN summit, is the goal toward quality education. However, of the 17 goals named for sustainable development that universally apply to all member countries, almost all dovetails with quality education as being increasingly integral to their vitality. At the center of the SDGs lies the effort to end all forms of poverty and fight inequalities globally. However, it is aptly expressed in the declaration statement that "obtaining a quality education is the foundation to creating sustainable development. In addition to improving quality of life, access to inclusive education can help equip locals with the tools required to develop innovative solutions to the world's greatest problems" (UNSED 2019). Nevertheless, not all potential young people are able to fulfill the dream of higher education. This is especially true in Nigeria. For instance, 28% and 30% of students who applied for university admission were admitted in 2010 and 2015, respectively (JAMB 2017). The female vs male admissions for the same years were 41.5% vs 58.50% in 2010 and 42.56% vs 57.44% in 2015. The difference in the admission shows a clear gender disparity. Some studies have raised the challenges of this disparity and its implications for national development (Oludayo et al. 2019; Onwuameze 2013; Salman et al. 2011; Kazeem et al. 2010; Lincove 2009; Adeyemi and Akpotu 2004; Aja-Okorie 2002; Adeyemi 2001).

Many studies have shown that gender disparity in education attainment has a huge impact on the participation of women in skilled employment and other human development strides. This is a major challenge in meeting the SDGs Goal 4 of "Achieving inclusive and quality education for all" by 2030. By this, the goal is aimed at eliminating gender disparities in education and ensuring equal access to all levels of education and vocational training for the vulnerable, including persons with disabilities, indigenous people, and children in vulnerable situations. Similarly, Section 18 of the 1999 Federal Republic of Nigeria's constitution dealing with the fundamental principles of state policy reflects the nation's commitments to equality of all, irrespective of race, sex, or gender. Therefore, the National Policy of Education (2004) stipulates that every Nigerian child should be given equal educational opportunity. Hence, there is a need for empirical evidence at sub-national level for assessing the gender disparity in higher education enrollment in Nigeria.

Recent evidence-based studies showed that there is gender inequality in higher education in Nigeria. For instance, Oludayo et al. (2019) found that males continuously dominate the admissions into higher institutions of learning between 2010 and 2015. With the growing interest in Sciences, Technology, Engineering and Management (STEM) disciplines, gender inequality has been documented in the enrollment system with lower female enrollment in STEM. For instance, Salman et al. (2011) found significant gender and spatial differences in the enrollment of mathematics education. Saint et al. (2003) observed that the share of science and engineering courses in the total enrollments increased from 54% in 1989 to 59% in 2000 and in favor of males. However, there are very few studies on gender disparity in education attainment in Nigeria using geospatial techniques.

Geographic Information Systems (GIS) technologies have been used by researchers, government agencies, and the commercial world for examining spatio-temporal relationships and patterns in a wide range of urban contexts (Haworth et al. 2013). However, research involving the use of GIS in examining issues associated with education such as education performance and infrastructure distribution has been relatively recent. Archibong et al. (2015) employed spatial statistical techniques to examine the spatial autocorrelation of power, sanitation, and water across 68,627 schools in 774 Local Government Areas in Nigeria and found evidence for clustering of nonfunctional infrastructure, aligned along Nigeria's six geo-political zones.

Kimosop et al. (2015) examined nationwide performance on the 2011 national Kenya Certificate of Primary Education (KCPE) exam by gender using Geographic Information Systems (GIS) and spatial statistics. Specifically, hot spot analysis was performed to see if unusual concentrations of scores exist in space in terms of overall performance, performance by gender, and individual subject scores. They found significant variability in terms of overall performance, performance by gender, and individual subject scores; with males outperforming females in mathematics, science, and social studies and religious education (SSR).

Similarly, Ansong et al. (2015) examined the spatio-temporal trends of academic performance at the junior high school level in Ghana since 2009 by using multilevel growth curve modeling, spatial statistics, and district level longitudinal data. Results revealed three statistically distinct trajectories of academic performance. The first statistically distinct trajectory reflects districts whose Basic Education Certificate Examination (BECE) results are neither improving nor declining consistently. The second statistically distinct trajectory refers to districts whose BECE pass rates have decreased consistently over time. The last statistically significant trajectory represents districts whose results increased steadily during the study period. Results also show that rural–urban gaps explained 31% of the performance trajectories.

In a recent study, Ansong et al. (2018) examined the spatial patterns of gender inequality in junior high school enrollment and the educational resource investments associated with the spatial trends in 170 districts in Ghana. The study used hot spot analysis based on the Getis-Ord G_i statistic, linear regression, and geographically weighted regression to assess spatial variability in gender parity in junior high school enrollment and its association with resource allocation. The results revealed rural–urban and north–south regional variability in gender parity. At the national level, educational expenditure, and the number of classrooms, teachers, and available writing places had the strongest positive associations with girls' enrollment. These relationships were spatially moderated, such that predominantly rural and northern districts experienced the most substantial benefits of educational investments.

This current analysis contributes to the growing discourses on gender disparity in education by applying spatial analysis at the state level in Nigeria for the period from 2005 to 2015 with the following objectives: (i) determine gender inequalities in university enrollment in Nigeria, (ii) identify spatial patterns and clustering of percentage female enrollment in Nigerian universities on a yearly basis between 2005 and 2015, (iii) examine the pattern of percentage female enrollment across the five disciplines common to all universities, and (iv) identify the factors that may be associated with the spatial patterns of gender inequality in university enrollment between 2005 and 2015 at state level.

Method

Study Area

Nigeria lies between latitudes 4°–14°N and longitudes 2°–15°E (figure 10.1). It has a total area of 923,768 km². The country has a federal system of government with 36 states and the Federal Capital Territory of Abuja (FCT). Within the states and the FCT, there are 744 local governments in total. Nigeria is

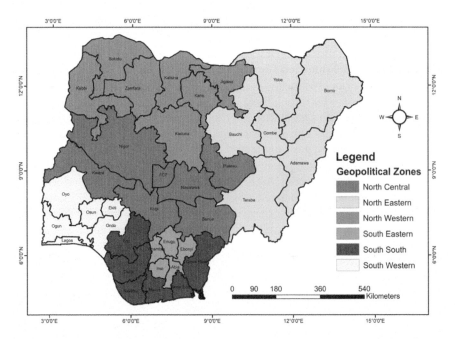

FIGURE 10.1
The location and geo-political zones of Nigeria.

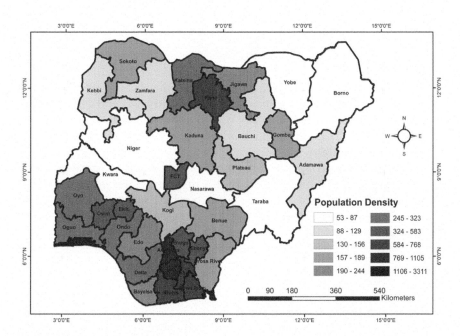

FIGURE 10.2
Population density of Nigeria in 2015 based on National Population Commission and National
Bureau of Statistics Estimates.

divided into six geo-political zones: north-central; northeast; northwest;
southeast; south-south, and southwest (Olawole and Olapoju 2018). The country's estimated population in 2015 was 181,137,448 with a population density
of 199 per km^2. Figure 10.2 shows the population density of Nigeria in 2015.

Nigeria has different natural resources in abundance. For instance, the
country is Africa's biggest oil exporter, and has the largest natural gas
reserves on the continent. Hence, Nigeria's growth performance is highly
influenced by oil price volatility. Between 2000 and 2014, Nigeria's gross
domestic product (GDP) grew at an average rate of 7% per year. The gross
domestic product (GDP) growth rate dropped to 2.7% in 2015. Since 2015, economic growth remains muted following the oil price collapse in 2014–2016
(World Bank 2019).

Agricultural growth in the country remains below potential due to continued insurgency in the northeast and ongoing farmer-herdsmen conflicts. The
decline of the agriculture sector weakens prospects for the rural poor, while
high food inflation adversely impacts the livelihoods of the urban poor.

Employment creation in Nigeria remains weak and insufficient to absorb
the fast-growing labor force, resulting in high rate of unemployment (23% in
2018), with another 20% of the labor force underemployed. Furthermore, the
instability in the north and the resulting displacement of people contribute
to the high incidence of poverty in the northeast (World Bank 2019).

Inequality in terms of income and opportunities has been growing rapidly and has adversely affected poverty reduction. The north–south divide has widened in recent years due to the Boko Haram insurgency and a lack of economic development in the northern part of the country. The lack of job opportunities is at the core of the high poverty levels, of regional inequality, and of social and political unrest in the country. Nigeria's human capital development also remains weak due to under-investment. The country is ranked 152 of 157 countries in World Bank's 2018 Human Capital Index (World Bank HCI 2018).

Nigeria as a secular nation is made up largely of highly developed and diversified religious groups. Apart from Christianity and Islam, which are the most widely practiced, there are numerous local faiths grouped together as traditional religions. Most Nigerian Muslims are Sunni and are located in the northern parts of the country (northwest, northeast, and parts of north-central) while the Christian population is located mainly in the part of north-central and the southern areas of the country (southeast; south-south, and southwest).

The development of modern education systems in Nigeria can be traced to the introduction of Western education. Western education refers to formal education introduced to the country by British missionaries in the 19th century. It began when the first primary school was established by the Methodist missionaries in 1843 in Badagry, Lagos. The flag-off of secondary school education commenced with the C.M.S Grammar School also in Lagos established in 1859. By 1914, when both the Northern and Southern Protectorates were amalgamated, there were 59 government and 91 mission primary schools, as well as 11 secondary schools in the southern part of the country (Fafunwa 1974).

The spread of Western education in the north was not as smooth as it was in the south. This was because the north had the Islamic system of education for many years before the introduction of Western education. While primary school enrollment rose from 66,000 students in the north in 1947, to about 206,000 students within 10 years, there was an astronomical increase from 240,000 to 1,209,000 in enrollment within the same period in the southern parts of the country. The number of secondary school students in the entire nation grew much less dramatically, increasing from 10,000 in 1947 to 36,000 in 1957 with almost 90% of this growth accounted for in the southern part.

In the 1950s, Nigeria adopted the British system of education and at Independence (1960), there were only six higher educational institutions in Nigeria: the University of Ibadan, the University of Ife, the University of Lagos, Ahmadu Bello University, the University of Nigeria at Nsukka, and the Institute of Technology at Benin (Nduka 1975). One of the six is located in the Northern Nigeria. The number of universities had increased to 174 by 2019 (NUC 2019).

Presently education is administered by the federal, state, and local governments. The Federal Ministry of Education is responsible for overall policy formulation and ensuring quality control but is primarily involved with tertiary education. Nigeria's education system encompasses three different sectors: basic education, post-basic/senior secondary education, and tertiary education. According to National Policy on Education (NPE 2004), basic education covers nine years of formal (compulsory) schooling consisting of six years of elementary and three years of junior secondary education. Post-basic education includes three years of senior secondary education. The secondary school certificate examination (SSCE) is taken at the end of the three years of senior secondary school education. Passing at least five subjects including English language and mathematics at credit level in SSCE is an essential requirement for university admission in the country.

The tertiary education covers four to six years, depending on the program of study. At the tertiary level, the system consists of a university sector and a non-university sector. The latter is composed of polytechnics and colleges of education (World Education News Review 2017).

The tertiary sector as a whole offers opportunities for undergraduate, graduate, vocational, and technical education. The National University Commission (NUC) is the government regulatory agency in charge of university administration in Nigeria. There are 174 approved universities in Nigeria: 43 federal owned, 52 owned by states, and 79 privately owned (NUC 2019). Entrance examination into universities in Nigeria is conducted by the Joint Admission and Matriculation Board (JAMB). It was established as a central body for the conduct of admissions into Nigerian universities. The admission policy officially recognizes four admission criteria namely: merit (40%); educationally disadvantaged states (20%); university discretion (10%); and catchment area (30%) (JAMB 1978). These criteria specifically are with the view of removing the gap in university enrollment among regions of the country. The board is also charged with the responsibility to administer similar examinations for applicants to Nigerian public universities and polytechnics.

Data

The study used secondary data. The first data set on university enrollment by gender for the years of 2005–2015 were extracted from the publications of the Nigerian National Bureau of Statistics (NBS 2012) and that of the Joint Admission and Matriculation Board (JAMB 2017). The study period was selected based on the availability of university enrollment data per state and per discipline. The second dataset consists of possible factors that may influence geographical variation in percentage female enrollment (PFE) in universities and was obtained from the Nigerian Annual Abstract of Statistics published by National Bureau of Statistics (NBS 2012, 2016). The data attributes include population size, number of local government area, number of secondary schools, number of students with at least five credit pass (including mathematics and English language) in senior secondary school certificate

TABLE 10.1

Data Types

Type	Content	Source
University admission data	2005–2015 total enrollment by gender per state	JAMB 2017
	2005–2015 total enrollment by discipline and gender per state	JAMB 2017
GIS data	Shape file of Nigerian boundary	Diva-GIS 2019
	Shape file of state boundaries	Diva-GIS 2019
Socio-economic statistical data	Total number of applicants to Joint Admission and Matriculation Board (JAMB) for admission to universities in Nigerians	
	Number of local government area per state	NBS 2012, 2016
	Number of public federal universities per state	JAMB 2017
	Number of public state universities per state	JAMB 2017
	Number of private universities per state	JAMB 2017
	Number of secondary school teachers per state	NBS 2012, 2016
	Number of senior secondary schools	NBS 2012, 2016
	Number of students with at least five credit pass in senior secondary school certificate in 2015	NBS 2012, 2016

examination and the number of federal, state, and privately owned universities in each state of the country (table 10.1).

The third dataset consists of spatially referenced data of the 36 states and Federal Capital Territory (FCT) boundaries. The spatial data on Nigeria were downloaded from the porter of DIVA-GIS (http://diva-gis.org/gdata).

Statistical and Spatial Analyses

To map and analyze the datasets, ArcGIS 10.4 software (ESRI Inc. 2016) was used. University enrollment data and percentage female enrollment (PFE) were merged with the spatially referenced data of the 36 states and FCT boundaries using the "Join table" feature in ArcGIS® software. The yearly percentage female enrollment (PFE) in universities was computed using the formula:

$$PFE = \frac{\text{Total Enrollment}}{\text{Female Enrollment}} \times 100 \tag{1}$$

When the PFE is greater than 50%, there is inequality in favor of females, and when PFE is less than 50%, the inequality favors males. When PFE is 50%, it denotes gender equality in enrollment. The computed yearly PFE were classified into four groups of equal intervals and used to generate Choropleth maps showing spatial pattern of percentage female enrollment in universities in the country for the study period (2005–2015).

Further, the following geospatial statistical tools were used in the study: Global Moran I (Moran 1948), Local Moran's I (Anselin 1995), and Getis and Ord (1992) Gi^*.

The Global Moran I statistic (Moran 1948), a measure of spatial autocorrelation, refers to the tendency that a value of a variable at a location is correlated to the values of the same variable at nearby locations Cliff and Ord (1973, 1981). It measures spatial autocorrelation based on both feature locations and features values simultaneously. The Global Moran's I statistic is expressed as follows:

$$I = \frac{N}{\sum_{i=1}^{N}\sum_{j=1}^{N} w_{ij}} \times \frac{\sum_{i=1}^{N}\sum_{j=1}^{N} w_{ij}\left(X_i - \bar{X}\right)\left(X_j - \bar{X}\right)}{\sum_{j=1}^{N}\left(X_i - \bar{X}\right)^2} \tag{2}$$

Where n is the number of state, X_i is the PFE associated with state i, X_j is the PFE at another state j (where $i \neq j$), $\bar{X}.$ is the mean of the PFE, and w_{ij} is a weight applied for comparison between state i and state j.

In general, the values of Global Moran's I would be between -1 and $+1$ (Goodchild 1986). Negative autocorrelation values (-1) mean nearby locations tended to have dissimilar values; positive autocorrelation values ($+1$) mean that similar values tended to occur in adjacent areas; and values close to zero indicate no spatial autocorrelation (Goodchild 1986). Along with the index, Z-scores are usually reported for the statistical significance test. If Z is out of ±1.96, the null hypothesis of the randomness test is rejected at the 95% confidence level, which means the pattern is spatially auto-correlated (Gao et al. 2016). The Global Moran's I was computed to identify clustering (spatial autocorrelation) in gender disparity in university enrollment in the country.

Spatial dependence is defined as the association between a value at a particular state and values of nearby states. Therefore, in order to further indicate the spatial dependence in gender disparity in university enrollment in the country, the Local Moran's I and Getis and Ord (1992) Gi^* statistics were computed (Celebioglu 2017; Djukpen 2012; Lee and Rogerson 2007; Cromley and McLafferty 2002).

Local Moran's I indicates how neighboring values are associated with each other. In other words, it identifies significant levels of spatial clustering at local levels. The formula expressed as follows:

$$I_j = \frac{\left(x_i - \bar{x}\right)}{\sum_i x_i - \bar{x}^2)} \sum_j W_{ij(x_i - \bar{x})} \tag{3}$$

Where $\bar{x}$ is the mean intensity over all observations (PFE), x_i is the intensity of observation in state i, x_j is the intensity for all other observations in state j (where $i \neq j$), and w_{ij} is a distance weight for the interaction between observations i and j.

Specifically, Local Moran's *I* analysis identifies five possible clustering patterns: High–High (HH) indicates clustering of high PFE (positive spatial autocorrelation), Low–High (LH) indicates that low PFE rates are adjacent to high ones (negative spatial autocorrelation). Low–Low (LL) signifies clustering of low PFE rates (negative spatial autocorrelation); High–Low (HL) indicates that high PFE values are adjacent to low PFE values (negative spatial autocorrelation), and not significant means no spatial autocorrelation.

The Getis and Ord (1992) *Gi** statistic detected units (states) with high or low concentrations of value of interest (PFE) within a region (Nigeria). The *Gi**statistic identifies either clustering of higher-than-average values (hot spots) or lower-than-average values (cold spots). Individual state and FCT in Nigeria were used as spatial units in this research, and data values were aggregated for each state (PFE). The standardized *Gi** statistic is:

$$G_i^* (d) = \sum_{i=1}^n W_{ij}(d) x_i \, / \sum_{i-1}^n x_i \; .... \qquad (4)$$

where x_i is the state *i*; W_{ij} is the spatial weight matrix between states *i* and *j*. If state *i* is adjacent to state *j*, then $W_{ij} = 1$; if state *i* is not adjacent to state *j*, then W_{ij} is 0. If *Gi** is significantly positive, then the value of states around *i* is relatively high (above average value). Accordingly, these states form a high-value cluster (hot spot area). If *Gi** is significantly negative, then the value of states around *i* is below the average value. Consequently, they form a low-value cluster (cold spot area).

Finally, ordinary least squares (OLS) regression and geographically weighted regression (GWR) were computed and compared to determine whether the direction and magnitude of the relationships between percentage female enrollment in Nigerian universities and the set of seven independent variables (possible determinants) varied from one spatial unit (state) to another.

ArcGIS 10.4 was used to implement both the OLS and GWR models. The Akaike's information criterion (AIC; Akaike 1974) was used to compare the OLS with the GWR. The AIC comparison reveals whether the spatial structure significantly enhances the model fit. In addition, variance inflation factor (VIF) procedure was used to test for multicollinearity. VIF values greater than 10 indicate multicollinearity, which reduces the explanatory power of the model (Monsour 2018; Cardozo et al. 2012; Menard 2002). Variables with a variance inflation factor (VIF) less than 10 were retained.

Results

Gender inequality in education attainment is an essential spatial and temporal event. The spatial and temporal dimensions of gender inequality in university enrollment in Nigeria between 2005 and 2015 are presented in the subsections below. Gender inequalities in university enrollment were examined in the section using the percentage of female enrollment (PFE).

Temporal Pattern of Percentage Female Enrollment in Nigerian Universities

Temporal distribution of male and female enrollment among the 36 states and the FCT is shown in table 10.2. The mean female enrollment in each of the years is lower than male enrollment for the same year. The mean female enrollment exhibits an increasing trend from 869.8 in 2005 to 5283.9 in 2013. The trend reduces to 4870.3 in 2014 and increase again to 5583.3 in 2015 (table 10.2).

The percentage female enrollment varied among states and from year to year. Table 10.3 shows the minimum and maximum distribution of PFE among the states on yearly basis. The lowest (13.83%) occurred in 2011 while 2015 recorded the highest PFE (55%). Similarly, the number of states based on the classification of PFE into four categories is also shown in table 10.3. The table clearly shows that states with above 50% female enrollment are very few. For instance, the PFE stayed below 50% of the total enrollment in all the states except in Anambra and Imo states (table 10.4). In 2012 and 2014, Anambra, Enugu, and Imo states, all three states in southeastern part of the country, out of the 36 states of the federation had gender disparity in favor of females (PFE > 50%).

Spatial (Clustering) Pattern of Percentage Female Enrollment in Nigerian Universities

A persistent south to north disparity in female enrollment is shown in figure 10.3. Between 2008 and 2015, states in the southern part of the country consisting of three zones (southwest, southeast, and south-south) have higher percentage of female enrollment in the range of 38–49.9% while states in the Northern Nigeria (northwestern, northeastern, and north-central) all had PFE below 38%, except Kaduna and Plateau states that had PFE between 38% and 49.99% in 2011–2015. The Federal Capital Territory (2005, 2018) and Borno (2009, 2015) had PFE between 38% and 49.99% (figure 10.3).

The overall spatial clustering of PFE and how it varied in different parts of the country were examined using the Moran's *I* Index. Table 10.5 shows the Global Moran's *I* Index values and their associated Z-scores and level of significance. The values indicate the existence of significant spatial clustering of PFE in the country. However, the degree of clustering slightly varied (Z-Scores) among the years studied (see Appendix 10.1).

Further analysis of the detailed spatial patterns with G-statistics was used to examine the degree of clustering of PFE at the state level (local level). The "hot spots" and "cold spots" of PFE in the study years are presented in figure 10.4.

TABLE 10.2

Distribution of Male and Female Enrollment, 2005–2015

Year	Male Enrollment				Female Enrollment			
	Minimum	Maximum	Sum	Mean	Minimum	Maximum	Sum	Mean
2005	51	4,878	45,924	1,241.2	11	5,189	32,184	869.8
2006	51	4,728	52,227	1,411.5	24	5,118	36,019	973.5
2007	110	4,987	70,961	1,917.9	51	4,983	47,095	1,272.8
2008	76	5,953	71,958	1,944.8	52	5,774	48,219	1,303.2
2009	131	6,865	91,790	2,480.8	67	5,979	59,774	1,615.5
2010	603	14,155	247,751	6,695.9	266	14,570	175,778	4,750.8
2011	659	12,785	241,003	6,513.6	289	13,907	176,338	4,765.9
2012	507	13,881	255,557	6,906.9	282	14,210	185,299	5,008.1
2013	717	13,721	267,891	7,240.3	37	323	16,179	5,283.9
2014	831	13,660	257,498	6,959.4	323	12,588	180,201	4,870.3
2015	1,160	16,201	278,756	7,533.9	565	15,501	206,582	5,583.3

TABLE 10.3

Temporal Trend in Percentage Female Enrollment in Nigerian Universities (2005–2015)

	Percentage of Female		Number of States			
Year	Minimum	Maximum	0–24.99%	25–37.99%	38–49.99%	50–72.99%
2005	17.19	52.21	5	20	10	2
2006	15.84	52.11	4	19	12	2
2007	17.29	51.04	6	15	15	1
2008	19.77	49.34	7	12	18	0
2009	16.81	49.36	8	12	17	0
2010	15.37	51.47	7	11	17	2
2011	13.83	53.43	5	11	19	2
2012	14.09	52.14	6	10	18	3
2013	14.50	54.11	7	9	17	4
2014	15.24	54.31	8	10	16	3
2015	18.41	55.00	6	8	19	4
2005–2015	15.90	52.63	7	10	18	2

TABLE 10.4

States with Percentage Female Enrollment Above 50%

Year	States with PFE in Favor of Females				Percent of States (%)
2005	Anambra	Imo			5.41
2006	Anambra	Imo			5.41
2007	Anambra				2.70
2008	None				0
2009	None				0
2010	Anambra	Imo			5.41
2011	Anambra	Imo			5.41
2012	Anambra	Imo	Enugu		8.10
2013	Anambra	Imo	Enugu	Abia	10.81
2014	Anambra	Imo	Enugu		8.10
2015	Anambra	Imo	Enugu	Abia	10.81

Hot spots refer to states with clustered high G-statistics values, which are also states with high percentage female enrollment. By contrast, cold spots are states with clustered low G-statistics values and low percentage female enrollment.

Pattern of Female Enrollment across Disciplines in Nigeria: 2011–2015

This section presents PFE in six faculties that are common to all universities in Nigeria. In Nigerian universities a faculty consists of: a group of university departments concerned with a major division of knowledge – e.g., Science, Social Sciences, Law, Engineering, Education, and Medicine. Descriptive

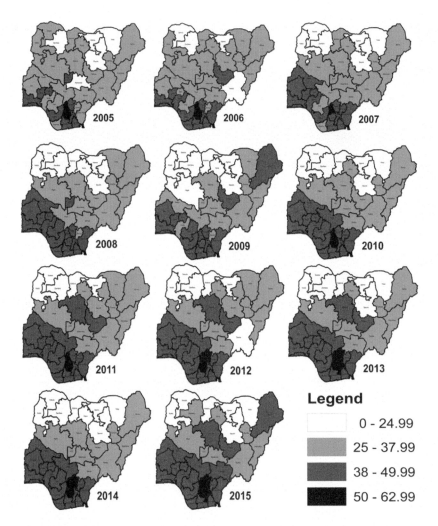

FIGURE 10.3
Spatial patterns of percentage female enrollment, 2005–2015.

statistics of the female vs male enrollment from 2011 to 2015 across the faculties is as presented in Appendix 10.2.

The pattern of female enrollment in six faculties in Nigeria universities between 2011 and 2015 across disciplines showed a high female enrollment in Education courses (over 50% in 17 of the 36 states and FCT). About 15 (40.54%) of the states had more females than males enrolled in Medical School while additional 17 states had between 34% and 49% of females enrolled in Medicine (table 10.6). In Social Sciences and Law courses, females had higher proportion in 4 states and almost 20% enrollment in 7 states of the country respectively. However, the two faculties – Social Sciences and Law – had

TABLE 10.5

Clustering Results of Percentage Female Enrollment in
Nigerian Universities, 2005–2015

Year	Value	Z Score	Significant*	Comment
2005	0.766	7.890	0.000	Clustered
2006	0.841	8.657	0.000	Clustered
2007	0.826	8.463	0.000	Clustered
2008	0.813	8.295	0.000	Clustered
2009	0.695	7.128	0.000	Clustered
2010	0.847	8.631	0.000	Clustered
2011	0.820	8.401	0.000	Clustered
2012	0.822	8.412	0.000	Clustered
2013	0.854	8.705	0.000	Clustered
2014	0.869	8.824	0.000	Clustered
2015	0.859	8.738	0.000	Clustered
2005–2015	0.854	8.700	0.000	Clustered

* Values in the column were approximated

more states with female enrollment between 34% and 49% categories than other faculties. Social Sciences had 51.35% in 19 states and Law had 75.67% in 28 states. No states had PFE above 50% in the faculties of Engineering and Sciences during the study years. In Engineering, PFE did not exceed 33% for all states while 26 (72.27%) of the states had PFE in the category of 34–49%.

Over 50% female enrollment in Education courses are all concentrated in the southern part of the country (Abia, Akwa Ibom, Anambra, Bayelsa, Cross River, Delta, Ebonyi, Edo, Ekiti, Enugu, Imo, Lagos, Ogun, Ondo, Osun, Oyo, and Rivers states). In terms of spatial distribution for female enrollment for Law School, there is a general persistent higher PFE rates across the states, except for Sokoto and Zamfara states. Seven states – Adamawa, Bauchi, Borno, Delta, Imo, Lagos, and Ogun – had over 50% in PFE in the Law School. Additional 28 states had PFE between 34% and 49%. The pattern of enrollment in degree programs in Social Sciences as shown in figure 10.5 revealed that PFE differed across the states. Only Abia, Anambra, Delta, and Imo had over 50% PFE during the study years. About 19 states had PFE in the range of 34–49%. The distribution is disproportionately skewed in favor of southern states (PFE above 50%) in enrollment in Medical School with 14 states, all concentrated in the southern region. Only Jigawa in the northern part of the country had a PFE in favor of females.

The gender disparity in enrollment with respect to Engineering courses shows fewer females than males enrolled in Engineering School across the states in the country. The highest PFE among the states is between 17% and 33%. Twenty-two states had between 17% and 33% while 15 states had less than16%.

In the Sciences, no state of federation had female enrollment more than male enrollment. The highest PFE in Science faculty is between 34% and 49% in 26 states spread across four regions while 12 states had between 17% and 33%.

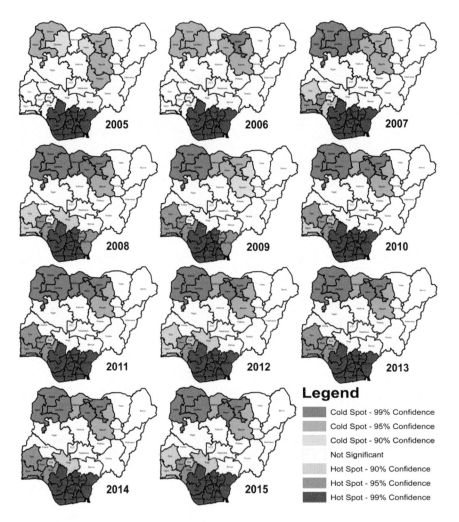

FIGURE 10.4
Hot and cold spots of PFE, 2005–2015.

Predictors of Spatial Pattern of Percentage Female Enrollment: 2005–2015

Results of Global Model

Table 10.7 provides the results of the OLS model and lists the eight predictors. The OLS model explained 68.18% of the variance in percentage female enrollment (Adjusted R^2 = 0.6818), and two of the eight predictors were significantly associated with PFE. The OLS regression model indicates that total

TABLE 10.6

Percentage Female Enrollment across Faculties, 2011–2015

| Faculty | Female Enrollment per State | | | | | | | | | | | | Total | |
| | 1–16% | | 17–33% | | 34–49% | | 50% and above | | | | | | | |
	states	Percent	states	Percent	states	Percent	states	Percent	States	Percent
Education	1	2.7	12	32.43	7	18.92	17	45.95	37	100
Social Sciences	1	2.7	13	35.14	19	51.35	4	10.81	37	100
Engineering	15	40.54	22	59.46	0	0.00	0	0.00	37	100
Law	0	0.00	2	05.41	28	75.67	7	18.92	37	100
Medicine	0	0.00	5	13.51	17	45.95	15	40.54	37	100
Sciences	0	0.00	11	29.73	26	70.27	0	0.00	37	100

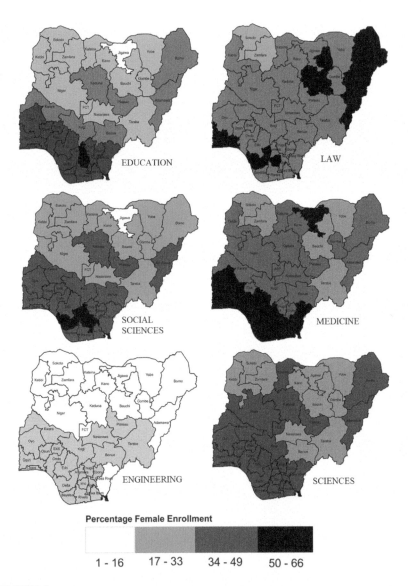

FIGURE 10.5
Percentage female enrollment across faculties.

application to universities ($b = 0.0032$, $p < 0.01$) and number of LGAs per state ($b = -0.438$, $p < 0.05$) were positively and negatively associated with PFE, respectively. The other six predictors were not significant. An examination of the VIF result revealed that multicollinearity did not bias the OLS estimations as the highest VIF value was 2.19 (table 10.6), which was well below the common cut point of 10 (Monsour 2018; Menard 2002); hence, the two variables were used in the GWR model specification.

TABLE 10.7

Summary of OLS Model Coefficients.

Variable	Estimate	Std-error	t-Statistic	p-value	VIF
Constant	35.002	3.892	8.99	0.0001*	
Total Application to Universities	0.003	0.001	6.79	0.0001*	2.186
Number of Local Government Areas	–0.483	0.173	–2.53	0.0174*	1.902
Total Secondary School Teachers	–0.001	0.002	0.00	0.9982	2.687
Total Students with 5 credits in SSCE including Mathematics and English	0.006	0.002	0.21	0.8324	2.198
Total Senior Secondary School Enrollment	–0.003	0.003	–0.88	0.3844	1.694
Number of State Universities	–1.483	1.693	–0.88	0.3883	1.233
Number of Federal Universities	0.434	2.688	0.16	0.8726	1.591
Number of Private Universities	0.269	0.534	0.50	0.6183	2.186

* Statistically significant at 0.01 level

Spatial Patterns of Local Estimates and Model Fit

To account for spatial variability in PFE, geographic weighted regression (GWR) was used. Comparing the two models as shown in table 10.8, the OLS model explains 68.18% of the total variance of the PFE across the states in Nigeria with an AICc of 251.27, while the GWR model shows a significant improvement over OLS based on two criteria. First, the GWR model increased the amount of variance explained by 18 points, yielding 86% variance explained. Second, the GWR model provided a better fit to the data

TABLE 10.8

Estimated Diagnostics of OLS and GWR Models

Variable	OLS	GWR
R^2	0.755	0.91
Adjusted R^2	0.682	0.863
AICc	251.273	219.852

Note: OLS = ordinary least squares; GWR = geographically weighted regression; AIC = Akaike information criterion.

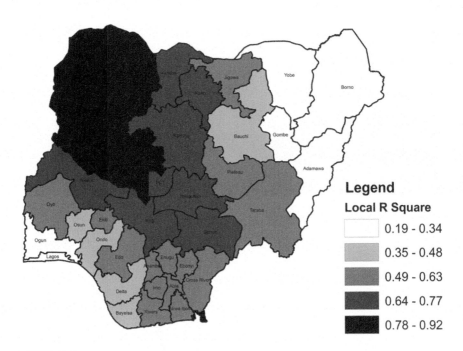

FIGURE 10.6
Spatial distribution of the local R^2 estimates across the study area.

because the Akaike information criterion score for the GWR model (291.852) was lower than the AICc score for the OLS model (251.273). The 31.421 points difference between the two AICc scores is well above the generally accepted cutoff of 3 points or greater (Jiang and Xu 2014; Ansong et al. 2015).

Figure 10.6 shows the local R^2 values across the 36 states and the FCT in Nigeria. As shown, the total variance explained by the local model ranges from 19% to 92%. This model fits the data well in many states. These are states where PFE falls below 37.99% (see figure 10.3). The model predicted over 50% in 14 states spread across the six geo-political zones in the country. The model does not fit the data well in the Ogun, Lagos, Osun, Delta, and Bayelsa states in the southwestern zone and Yobe, Borno, Gombe, and Adamawa in the northeastern part of the country. The adequacy of the final GWR model is further verified by the Moran's I test result: $M(i) = -0.069$, Z Score $= -0.4132$, $p = 0.67$.

Figure 10.7 shows the spatial distribution of the GWR parameter estimates for each explanatory variable. Figure 10.7A displays the spatial distribution of the model intercept. The positive values are found in the southern part of the country, where PFE are higher, mostly in the range of 38–49.99%, and 50% and above.

Figure 10.7B shows the spatial association between number of total applications received for university admission and percentage female enrollment.

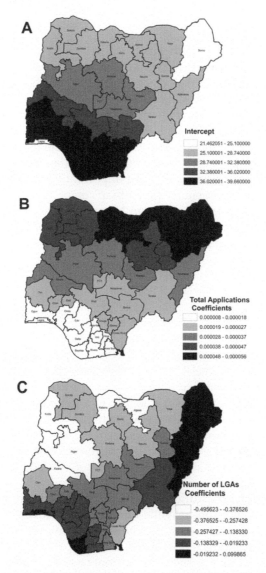

FIGURE 10.7
The geographically weighted regression parameter estimates for (A) the model intercept, (B) total application to universities, and (C) number of Local Government Areas per state.

The effect of the predictor, total applications to universities in explaining the variance of percentage female enrollment is very strong in the five northern states of Katsina, Kano, Jigawa, Yobe, and Borno. Figure 10.7C shows the association between number of LGAs and percentage female enrollment. The effect of the number of LGAs in explaining the variance of percentage female enrollment is very weak in the northwestern and north-central states.

Discussion

One of the most important things to have ever happened to Nigeria was education. Well before advent of the colonialist on the African continent, informal learning had become well established. Children, male and female, were taught the rich African culture and customs. Nigeria, like many other African regions had institutionalized roles and duties that cut across all ages and gender. While in many parts of the country the male folks dominated religious, social, and family spheres, the women largely held sway in commerce, local administration, and, most importantly, in child training. These roles were more complimentary rather than competitive.

The coming of formal education in Nigeria, however, tilted this balance more in favor of the males. Right from the inception of schools, particularly the primary level, female enrollment had taken the back bench for many obvious male dominating reasons. At the post-primary and secondary levels, the situation had almost remained a pattern. However, there exists a significant geographic variation in female enrollment across the country, particularly at the university level of education. Similar regional inequality in the PFE rates has also been observed in other countries of West, Central, and North Africa, and South Asia (Filmer 2000).

In Nigeria, female enrollments and educational outcomes are generally higher in the three southern geo-political zones than in the three northern zones. This is in consonance with previous studies of Kazeem et al. (2010) and Lincove (2009) that holds that female enrollment in the more Christian dominated southern states is more likely higher than the Islam-dominated northern part of Nigeria.

The reason for the regional gap is provided in an earlier study by Adeyemi (2001) which suggests that the diverse level of educational awareness and attainment among the various groups and zones in the country could be attributed to the fact that different parts of the country encountered Western education at different times. While formal education had its incursion and acceptance into the southern part of Nigeria, specifically Lagos and Calabar in the 1840s, the more Islam-dominated northern part did not accept "Western styled" education, as it was seen as an affront on their religious and socio-cultural lives. By the time Western education had become well established in the south, the north was just coming to terms with the globalization of education.

A closer look at female enrollment in Nigeria universities further shows that while there is gender imbalance in university education enrollment from the northern to southern parts of the country in general, regional and inter-state inequalities can still be observed. For instance, while it has been established that female enrollment in universities in the southern part of Nigeria is higher than what is obtainable at the northern part, a number of states in the southern part, particularly the south-south, still have low female enrollment in Nigerian universities. The situation in the southeastern states, notably Anambra and Imo, is a bit different as there is more female enrollment in the universities in studied disciplines of Science, Social Sciences, Law, Engineering, Education, and Medicine. A plausible explanation for this is the drive for commerce among young males in the southeastern part of Nigeria. Also, the need to significantly improve on the social status of women in the southeastern part of the country, as against what used to be the prevalent custom some years back, could be another probable explanation for the increase in female enrollment in higher education.

As noted by Egboh (1973), a female child in the east was seen as a second-class citizen whose only moral obligation was procreation. To avert this and be able to have a voice in the highly male-dominated society, women saw education, particularly at the higher levels, as a potent weapon to deliver them from the strong hold of men's economic and social domination. Also, the economic value of female children in the southeastern part of Nigeria may be another plausible explanation for observed increases in female enrollment in Nigeria universities as the bride price demanded by bride family tends to higher with highly educated female children during marriage ceremony.

Aside from general north–south and interstate variations in female enrollment in Nigerian universities, there is also a gender imbalance in science technology engineering and mathematic (STEM) courses in Nigeria. Embracing STEM has direct implications for socio-economic development. As can be observed, there is great apathy in the percentage of female enrollment toward STEM in Nigerian universities. This situation has largely remained the same when compared with previous findings in Nigeria made by Onwuameze (2013) as well as Adeyemi and Akpotu (2004). The low level of PFE in Science and Engineering has been attributed to the lack of interest in Sciences and Mathematics at secondary education (Adeyemi and Akpotu 2004). While female enrollment in Medicine is higher in the southern parts of the country, the northern part makes up for this in Law, as there is a higher number of female enrollment in the north, notably in Bauchi, Borno, and Adamawa. This may be due to several reasons. First, it may be due to higher enrollment in the program by students from the southern part of the country who are qualified for admission based on merit criterion. Second, these states are predominantly Islamic but the fact that Christianity and traditional religions are widely practiced in some Local Government Areas of the states could have accounted for the interest in law degree. The third reason is that Islamic/Sharia law is offered only in the University of Maiduguri, in Borno, which may account for the higher PFE in the Law program in the state.

Conclusions and Recommendations

This chapter analyzed the spatial and temporal patterns of gender enrollment in Nigerian universities. From the analysis, it can be concluded that there exists a wide gender gap in university enrollment between the geopolitical zones in Southern and Northern Nigeria.

The hot and cold spots on the maps indicate that the southwest, southeast, and south-south geo-political zones are hot spots of high percentage female enrollment while the northwest, northeast, and north-central zones had uninterrupted cold spots of PFE in the country. Specifically, the gender gap in university enrollment is wider in two of the three northern geo-political zones (northwest and northeast) than in the southern zones of the country. These northern zones experienced significant cold spots at different levels throughout the study years.

This enabled this analysis to conclude that females in the southern zones tend to have higher enrollment rates than those in the northern zones. Similarly, there are variations in the trend and pattern of female enrollment in the universities across the six geo-political zones, with wider gender gap existing in the northern parts of the country in enrollment in the faculties of Engineering, Medicine, and Education. The implication of this is that the present admission policy, aimed at reducing the enrollment gap between the southern and the northern parts in the country, did not account for how to bridge the gender gap in different disciplines. It only recognizes four admission criteria: merit (40%); educationally disadvantaged states (20%); university discretion (10%); and catchment area (30%) (Adeyemi 2004). The merit criterion clearly admits students who meet up with the admission requirements as fixed by the universities and the matriculation or admission board. The educationally disadvantaged states quota was developed to ensure that states with low student enrollment are adequately taken care of in the university admission processes. The catchment area quota, on the other hand, was enforced to take care of students within the service areas or threshold of the universities. This was used as a development strategy to ensure that university admission is not restricted only to the immediate constituency or geographical vicinities of the universities.

Consequently, reducing the gender gap among disciplines requires specifying gender criteria in higher education admission policy. This can be achieved by adopting a quota system which favors females in university admissions in Nigeria. Such special intervention would reduce gender gap in enrollment and eventually bring about gender balance in enrollment among the geo-political zones and in the various disciplines. This would increase the human resources frontier for the nation's social and economic development, since it is believed that when a woman is educated, the nation grows. This study will conclude by advocating the revision of the country's university admission criteria so as to accommodate provisions for the percentage of females to be admitted nationwide and also across disciplines. Also, the admission criteria should include the allocation of seats for female enrollment to qualified female applicants. As a proactive measure, state and local government officials of areas where primary and secondary education is not free should provide free education for females, especially in the northern zones. This type of policy could reduce the gender gap in enrollment across the different geo-political zones in the country.

Bibliography

Adeyemi, J.K. 2001. Equality of access and catchment area factor in university admissions in Nigeria. *Higher Education* 42(3): 307–332.

Adeyemi, K. and Akpotu, N. 2004. Gender analysis of student enrollment in Nigerian universities. *Higher Education* 48: 361–378.

Aja-Okorie, U. 2002. *Factors Associated with Gender Disparity in Enrollment Patterns in Nigeria Universities*. A PhD diss., University of Nigeria, Nsukka, Nigeria.

Akaike H. 1974. A New Look at the Statistical Model Identification . *IEEE transactions on Automalic Conlrol* 19, 716–723.

Anselin, L. 1988. *Spatial Econometrics: Methods and Models*. Kluwer Academic, Dordrecht.

Anselin, L. 1995. Local indicators of spatial association. *Geographical Analysis* 27(2): 93–115.

Anselin, L. 2002. Under the hood: issues in the specification and interpretation of spatial regression models.*Agricultural Economics* 27(3): 247–267.

Ansong, D., Ansong, E.K., Abena, O.A. and Afranie, S. 2015. A spatio-temporal analysis of academic performance at the basic education certificate examination in Ghana. *Applied Geography* 65: 1–12.

Ansong, D., Renwick, C.B., Okumu, O., Ansong, E. and Wabwire, C.J. 2018. Gendered geographical inequalities in junior high school enrollment: do infrastructure, human, and financial resources matter? *Journal of Economic Studies* 45(2): 411–425,

Archibong, B., Modi, V. and Sherpa, S. 2015. Geography of infrastructure functionality at schools in Nigeria: evidence from spatial data analysis across local government areas.*Papers in Applied Geography* 1(2): 176–183.

Cardozo, O.D., Garcia-Palomares, J.C. and Gutierrez, J. 2012. Application of geographically weighted regression to the direct forecasting of transit ridership at the station-level. *Applied Geography* 34: 548–558.

Celebioglu, F. 2017. Women employment in terms of gender inequality across the provinces of Turkey. *Eurasian Journal of Business and Economics* 10(19): 61–80.

Cliff, A.D. and Ord, J.K. 1973. *Spatial Autocorrelation.*Pion, London.

Cliff, A.D. and Ord, J.K. 1981. *Spatial Process: Models and Applications*. Pion, London.

Constitution of the Federal Republic of Nigeria, Federal Republic of Nigeria – 1999 - Federal Government Press Lagos.

Coombs, P.H. and Hallak, J. 1987. *Cost Analysis in Education: A Tool for Policy and Planning*. Johns Hopkins University Press, London.

Cromley, E. K., and McLafferty, S. L. (2002). Analyzing spatial clustering of health events. In E. K. Cromley and S. L. McLafferty (Eds.), GIS and public health (pp. 130–157).New York: Guilford Press.

Diva-GIS. *Nigeria Level Data*. Accessed August 10, 2019. http://diva-gis.org/gdata

Djukpen, R. O. 2010. Mapping the HIV/AIDS epidemic in Nigeria using exploratory spatial data analysis. *GeoJournal*, 77(4), 555–569.

Egboh, E.O. 1973. The place of women in the IIbo society of South-Eastern Nigeria, from the earliest times to the present. *Civilisations* 23/24: 305–16. Accessed January 8, 2020. http://www.jstor.org/stable/41229521.

Environmental Systems Research Institute Inc. (ESRI). 2016. *ArcGIS Desktop: Release 10.4*. Redland, California.

Esomonu, N.P.M and Adirika, B.N. 2012. Assessment of access to university education in Nigeria. *Research Journal in Organizational Psychology & Educational Studies* 1(5): 303–306.

Fafunwa, A.B. 1974.*History of Education in Nigeria*. George Allen& Unwin, London.

Federal Government of Nigeria. 1999. *The Constitution of the Federal Republic of Nigeria, Federal Government Press Lagos*.

Federal Republic of Nigeria. 2004. *National Policy on Education*. NERDC Press, Lagos.

Filmer, D. 2000. *The Structure of Social Disparities in Education: Gender and Wealth*. World Bank Policy Research Working Paper, 2268.

Fotheringham, A.S., Brunsdon, C. and Charlton, M. 2002. *Geographically Weighted Regression: The Analysis of Spatially Varying Relationships*. Wiley & Sons, Chichester.

Fotheringham, A.S., Charlton, M. and Brunsdon, C. 1997. Measuring spatial variations in relationships with geographically weighted regression'. In *Recent Developments in Spatial Analysis, Spatial Statistics, Behavioral Modeling and Computational Intelligence*, ed. M.M. Fischer and A. Getis, 60–83, Springer-Verlag, London.

Gao, Y., He, Q., Liu, Y., Zhang, L., Wang, H. and Cai, E. 2016. Imbalance in spatial accessibility to primary and secondary schools in China: guidance for education sustainability. *Sustainability* 8: 1–16.

Getis, A. and Ord, J. 1992. The analysis of spatial association by the use of distance statistics. *Geographical Analysis* 24(3): 189–206.

Goodchild M.F. 1986 Spatial autocorrelation. Concepts and Techniques in Modern Geography. Catmog 47, Geo Books. 1986, Geo Books, Norwich.

Haworth, B, Bruce, E. and Iveson, K. 2013. Spatio-temporal analysis of graffiti occurrence in an inner-city urban environment. *Applied Geography* 38: 53–63.

Joint Admissions and Matriculation Board (JAMB). 1978. Guidelines for Admission to First Degree Courses in Nigerian Universities. JAMB Lagos.

Kazeem, A., Jensen, L. and Stokes, S. 2010. School attendance in Nigeria: understanding the impact and intersection of gender, urban-rural residence, and socioeconomic status. *Comparative Education Review* 54(2): 295–319.

Kimosop, P.K., Otiso, K.M. and Ye, X. 2015. Spatial and gender inequality in the Kenya certificate of primary education examination results. *Applied Geography* 62: 44–61.

Lee, G., and Rogerson, P. 2007. Monitoring global spatial statistics. *Stochastic Environmental Research and Risk Assessment*, 21(5), 545–553.

Lee, K.H. and Schuett, M.A. 2014. Exploring spatial variations in the relationships between residents' recreation demand and associated factors: a case study in Texas', *Applied Geography* 53: 213–222.

Lincove, J.A. 2009. Determinants of schooling for boys and girls in Nigeria under a policy of free primary education. *Economics of Education Review* 28(4):474–484.

Mansour, S. 2018. Spatial patterns of female labor force participation in Oman: a GIS-based modeling. *The Professional Geographer* 70 (4), 593–608.

Menard, S. 2002. Applied logistic regression analysis. London: Sage.

Montgomery, D., Peck, E. and Vining, G. 2012. *Introduction to Linear Regression Analysis*. 2nd ed. New York: Wiley.

Moran, P. A. P. 1948. The Interpretation of Statistical Maps. *Journal of the Royal Statistical Society. Series B (Methodological)* , 10,243–251.

National Bureau of Statistics. 2012. *Annual Abstract of Statistics,* Accessed March 15, 2018. https://nigerianstat.gov.ng/download/253.

National Bureau of Statistics. 2015. *National Population Estimates.* Accessed October 15,2019. https://nigerianstat.gov.ng/download/474.

National Bureau of Statistics /Joint Admissions and Matriculation Board. 2017. *JAMB Admitted Candidates by State and Gender Within Faculty(2010–2016).* National Bureau of Statistics Abuja.

National Bureau of Statistics. 2016. Annual Abstract of Statistics, Accessed March 15, 2018. https://www.proshareng.com/admin/upload/report/ANNUALAB STRACTSTATISTICSVOLUME1.pdf.

National Universities Commission. 2019. *Universities in Nigeria.* Accessed June 10,2019. http://nuc.edu.ng/nigerian-universities/private-Universities.

Nduka, O.A. 1975. *Western Education and the Nigerian Cultural Background.* Ibadan: Oxford University Press.

Olawole, M.O. and Olapoju, O.M. 2018. Understanding the spatial patterns of tanker accidents in Nigeria using geographically weighted regression. *International Journal of Vehicle Safety* 10(1): 58–77.

Oludayo, O.A., Popoola, S.I., Akanbi, C.O. and Atayero, A.A. 2019. Gender disparity in admissions into tertiary institutions: empirical evidence from Nigerian data (2010–2015). *Data in Brief* 22: 920–933.

Onwuameze, N.C. 2013. *Educational Opportunity and Inequality in Nigeria: Assessing Social Background, Gender and Regional Effects.* A PhD diss., University of Iowa.

Saint, W., Hartnett, T.A. and Strassner, E. 2003. Higher education in Nigeria: a status report. *Higher Education Policy* 16: 259–281.

Salman, M.F., Yahaya, L.A. and Adewara, A.A. 2011. Mathematics Education in Nigeria. *International Journal of Educational Sciences* 3(1): 15–20.

Senadza, B. 2012. Education inequality in Ghana: gender and spatial dimensions. *Journal of Economic Studies* 39(6): 724–739.

Tesema, M.T. and Braeken, J. 2018. Regional inequalities and gender differences in academic achievement as a function of educational opportunities: evidence from Ethiopia. *International Journal of Educational Development* 60: 51–59.

United Nations Sustainable Development Goals. 2019. *Quality Education.* Accessed August 1, 2019. https://www.un.org/sustainabledevelopment/education/.

Utulu, C.C. 2001. Quality of university education in Nigeria: problems and solutions. *International Studies in Educational Administration* 29(1): 58–66.

World Bank. 2018. *Human Capital Index.* Accessed November 3, 2019. http://databank .worldbank.org/data/download/hci/HCI_2pager_NGA.pdf.

World Bank. 2018. *Human Capital Project.* World Bank, Washington, DC. License: CC BY 3.0 IGO. Accessed November 3, 2019. https://openknowledge.worldban k.org/handle/10986/30498.

World Bank. 2019. *Nigeria Overview.* Accessed November 3, 2019. https://www.wor ldbank.org/en/country/nigeria/overview.

World Education News Review. 2017. *Education in Nigeria.* Accessed November 3, 2019. https://wenr.wes.org/2017/03/education-in-nigeria.

World Population Review. 2019. Accessed November 3, 2019. http://worldpopulati onreview.com/countries/nigeria-population/.

Yahaya, L.A. 2004. Disparity in the enrollment of male and female undergraduates in science and technology based faculties at the university of Ilorin: implications for counseling. *Nigerian Journal of Counseling and Applied Psychology* 2(1): 186–201.

Appendix 10.1

Local Moran *I* Results of Percentage Female Enrollment in Universities in Nigeria, 2005–2015

Year	State	Value	Z-score	*p* Value	
2005	Abia	45.74	4.7279	0.0002**	HH
	Anambra	52.21	6.7219	0.0001**	HH
	Akwa Ibom	46.1	4.2636	0.0000**	HH
	Rivers	45.19	3.917	0.0001**	HH
	Imo	51.54	6.6707	0.0000**	HH
	Enugu	46.54	4.0499	0.0001**	HH
	Bauchi	24.73	2.1215	0.0339*	LL
2010	Abia	48.29	3.9045	0.0001**	HH
	Anambra	51.47	5.1538	0.0000**	HH
	Akwa Ibom	44.23	2.4645	0.0137*	HH
	Rivers	46.01	2.8462	0.0044**	HH
	Imo	50.72	4.8996	0.0001**	HH
	Edo	43.04	2.1363	0.0326*	HH
	Enugu	46.85	3.1995	0.0014**	HH
	Delta	45.35	3.014	0.0026**	HH
	Jigawa	15.37	3.4996	0.0047**	LL
	Kano	25.12	2.8803	0.0040**	LL
	Kebbi	21.42	3.3987	0.0007**	LL
	Sokoto	18.13	3.6291	0.0003**	LL
	Zamfara	23.93	3.5045	0.0005**	LL
	Bauchi	24.54	2.4394	0.0147*	LL
	Katsina	23.05	2.2778	0.0227*	LL
2015	Abia	51.52	3.7319	0.0002**	HH
	Anambra	55	0.0001	5.1463	HH
	Akwa Ibom	49.55	2.8304	0.0046**	HH
	Rivers	47.71	2.2391	0.0251*	HH
	Imo	53.67	4.5725	0.0005**	HH
	Edo	48.35	2.5286	0.0115*	HH
	Enugu	52.26	3.7701	0.0016**	HH
	Delta	48.6	2.7847	0.0005**	HH
	Jigawa	18.14	3.7316	0.0002**	LL
	Kano	26.73	3.5196	0.0004**	LL
	Kebbi	24.29	3.0858	0.0020**	LL
	Sokoto	22.62	3.2044	0.0014**	LL
	Zamfara	25.67	3.7521	0.0002**	LL
	Bauchi	24.78	2.6412	0.0083*	LL
	Katsina	21.74	3.0682	0.0022**	LL

Year	State	Value	Z-score	*p* Value	
2005–2015	Abia	48.4	3.662	0.0003**	HH
	Anambra	52.63	5.191	0.0000**	HH
	Akwa Ibom	45.59	2.5198	0.0117*	HH
	Rivers	45.96	2.5553	0.0169*	HH
	Imo	51.66	4.8039	0.0000**	HH
	Edo	43.86	2.0402	0.0413*	HH
	Enugu	48.74	3.4417	0.0006**	HH
	Delta	45.47	2.7161	0.0066*	HH
	Jigawa	15.9	3.6406	0.0003**	LL
	Kano	26.05	3.0257	0.0025**	LL
	Kebbi	22.17	3.5003	0.0005**	LL
	Sokoto	2012	3.7234	0.0002**	LL
	Zamfara	21.53	4.348	0.0000**	LL
	Bauchi	23.69	2.5541	0.0106*	LL
	Katsina	22.07	2.8022	0.0051**	LL

* Significant at the 0.05 level (2-tailed).
** Significant at the 0.01 level (2-tailed).

Appendix 10.2

Descriptive Statistics of Female vs Male Enrollment across Faculties, 2011–2015

Faculty	Year	Male				Female			
		Sum	Mean	Maximum	Minimum	Sum	Mean	Maximum	Minimum
Social Sciences	2011	29441	795.70	2871	108	21128	571.03	1683	28
	2012	40696	1099.89	2625	87	31871	861.38	2290	33
	2013	33273	899.27	2269	89	23825	643.92	1800	35
	2014	28957	782.62	2005	89	20154	544.70	1575	38
	2015	31593	853.86	1880	185	23613	638.19	1756	52
Education	2011	22894	618.76	1548	29	24159	652.95	2175	10
	2012	24271	655.97	1864	34	25731	695.43	2873	9
	2013	28722	776.27	2005	43	28700	775.68	2781	16
	2014	28575	772.30	1923	68	27538	744.27	2188	32
	2015	31995	864.73	2350	109	34901	943.27	3968	58
Engineering	2011	13966	377.46	956	26	3490	94.32	293	2
	2012	14774	399.30	1059	27	3831	103.54	340	6
	2013	19018	514.00	1388	32	4471	120.84	482	3
	2014	21864	590.92	1739	69	4940	133.51	429	6
	2015	23015	622.03	1991	59	4799	129.70	545	4

Faculty	Year	Male				Female			
		Sum	Mean	Maximum	Minimum	Sum	Mean	Maximum	Minimum
Law	2011	5307	143.43	347	14	4451	120.30	394	9
	2012	3707	100.19	272	7	3261	88.14	279	4
	2013	5165	139.59	371	16	4151	112.19	312	11
	2014	3491	94.35	239	9	3901	105.43	213	11
	2015	3243	87.65	211	11	3349	90.51	275	8
Medicine	2011	7763	209.81	680	17	8973	242.51	1260	7
	2012	6395	172.84	724	12	7335	198.24	1117	8
	2013	10359	279.97	798	1	12265	331.49	1597	16
	2014	8986	242.86	662	22	9554	258.22	1025	16
	2015	9715	262.57	798	20	10868	293.73	1017	20
Science	2011	35199	951.32	2040	73	21830	590.00	1749	43
	2012	72397	1956.68	4729	134	45403	1227.11	2870	68
	2013	36727	992.62	2283	66	22791	615.97	2022	33
	2014	36668	991.03	2035	55	22431	606.24	1668	32
	2015	40496	1094.49	2658	101	31980	864.32	4841	52

Index

Note: Page numbers in *italics* refer to figures.